D1103797

ENGINEERING MANAGEMENT

PEOPLE AND PROJECTS

A guide for the first level engineering manager, supervisor, or leader, as well as for the "manager without authority"—the project engineer, task leader, or lead engineer

Murray J. Shainis
Anton K. Dekom
Charles R. McVinney

BATTELLE PRESS
Columbus • Richland

658.5
Sh14e

Library of Congress Cataloging-in-Publication Data

Shainis, Murray J., 1926—
Engineering management: people and projects / by M.J. Shainis, A.K. Dekom, C.R. McVinney.

p. cm.
ISBN 0-935470-71-9 : $24.95
Includes index.
1. Engineering—Management. I. Dekom, Anton K., 1926–1994. II. McVinney, C. (Charles), 1945— . III. Title.
TA190.S48 1994 94-9155
658.5—dc20 CIP

Printed in the United States of America.

Copyright © 1995 Battelle Memorial Institute.
All rights reserved. No part of this book may be reproduced or transmitted in any form or by any means, electronic or mechanical, including photocopying, recording, or by any information storage and retrieval system, without written permission from the publisher.

Battelle Press
505 King Avenue
Columbus, Ohio 43201
Telephone 614-424-6393, 1-800-451-3543
Fax 614-424-3819

FOREWORD

Many books have been written on the subject of Engineering Management, and undoubtedly, there will be many more. Some have been on the people aspects, others on the project management aspects. Some have been written by psychologist types focusing on interpersonal interactions and guiding creative people, others by management scientists emphasizing quantitative methods and operations research techniques. Still others have emphasized the computer. Many have emphasized large, defense-oriented projects and the high-level project manager.

In this volume, we have written a book for the working engineering manager and project leader. You spend part of your time doing technical work, part of your time managing or supervising the work of others. You may not be formally recognized as a manager on the organization chart, but you have responsibility for one or more projects. Whether you are called a manager, section head, group leader, project manager/engineer/leader, lead engineer, or something else, this book is for you.

It is written in simple language and without references to academic tomes in non-engineering fields. It's approach is pragmatic. It is not a textbook: there are no cases, questions, exercises, or instructor's manuals. It offers a practical approach for technical people who have managerial or supervisory (and we don't quibble over the line between those two terms) responsibilities and who may not have received any formal training in that role.

Together, we have more than 110 years of relevant experience, both as managers and as consultants. We have presented training seminars to more than 25,000 engineers, scientists, and other

technical personnel throughout the United States, Canada, Europe, and Asia.

The book is divided in accordance with its title: "people" are the focus of Chapters 1 through 14, "projects" are discussed in Chapters 15 through 24. Obviously there are overlaps and connections.

And obviously, many people have been of great assistance in the completion of this book. Our publisher, Joe Sheldrick, and our editor, Cher Paul, have been of invaluable help with advice and ideas. Typists include Victoria Wright and Carol McNamara. Professional colleagues who contributed ideas, reviewed chapters, and provided a sounding board include Jack Davis, Richard Michaels, Nancy Lust, Dr. Linda Pollock-McVinney, and Janet W. Gagliano.

We wish our readers success in their careers and hope that we have been of some assistance to them.

—Murray Shainis
Tony Dekom
Chuck McVinney

CONTENTS

Tables

Figures

1

CHAPTER

MANAGING IN A TECHNICAL ENVIRONMENT: AN OVERVIEW

Let us start by examining the functions of a manager in the technical environment. What really is the job of an engineer who has been given the responsibility to manage a project and the project team? For our discussion, we'll use the model that was first developed during World War I by Henri Fayol in his classic work, *General and Industrial Administration.* Shown in Table 1-1, this remains, arguably, the primary teaching tool for understanding the role of the engineer as manager.

Planning, organizing, staffing, directing, and controlling. Some readers may say, "Gee. Isn't there a lot more to management than that?" In the model that we're using, these are the functions of the manager; other things—motivation, coordination, performance appraisal, for example—are the tools and methods used in executing these functions. What we hope to do in this volume is provide you with the tools, both diagnostic and prescriptive, that you will use in successfully implementing these five basic functions.

TABLE 1-1. Functions of a Manager

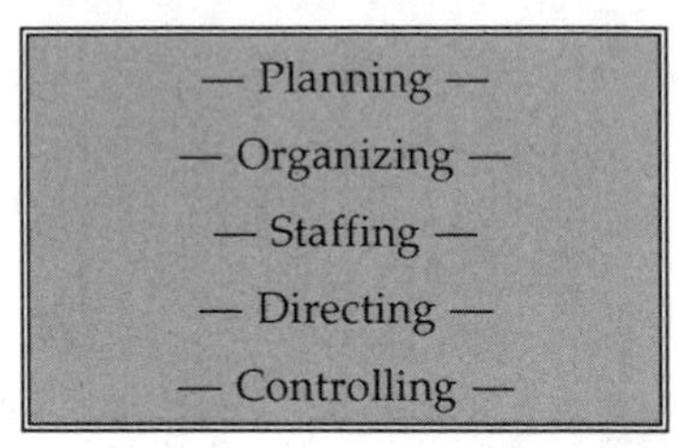

— Planning —

— Organizing —

— Staffing —

— Directing —

— Controlling —

PLANNING

Planning can be broken into four basic steps. At this point in our considerations, we are talking about planning from the managerial viewpoint; in later chapters, we will talk about the techniques of planning.

Objectives

The first part of planning is developing the goals, objectives, scope, and project definition that define for us what we want to do. This starts with a blank piece of paper. In some cases—in a revolutionary project, a major breakthrough project—this part of planning will be subject to continuing change well into the project. In certain other projects you are given a specification that someone else has developed, and that is your clear-cut goal. Assuming there are no problems, ambiguities, or errors in the specification, you have that clear-cut goal from beginning to end.

In most engineering organizations, the format of these goals will be twofold. First is the specification—or work statement, or detailed contract, but the "spec" is probably its most common name. Second is problem solution, where a problem surfaces and you're asked to solve it. It is important to understand from the beginning that problem solution presents a great obstacle in planning the scheduled projects, because often problem solution tasks come up quickly and take the form of unexpected surprises. You're now asked to bump other tasks to work on solving this problem, thus causing scheduling problems. One of the things that engineering managers must determine early in a project is how much time to allow for these unexpected surprises, these tasks not contained in the scheduled projects, and use that information in planning their schedules.

Work Breakdown Structure

The second part of planning is the work breakdown structure. Here we break down the work into increasingly smaller pieces and create what amounts to a "gazinta" diagram: this "gazinta" this "gazinta" this. A top-down breakdown, or bill of material, of a piece of hardware is a specialized work breakdown structure, but remember that we must have a work breakdown structure for the entire project, including project management, hardware, software, documentation, installation, training, special tools, test equipment, and so forth. In other words, the work breakdown structure must include all tasks necessary to complete the entire project. The format can be graphic, lists, or any of various other types. The graphic is generally the most popular, as we'll see in later chapters.

Scheduling

Scheduling sets *task vs time.* The two most popular forms of schedules are the bar chart (often called the Gantt chart after its inventor) and the network diagram. The network diagram, essentially a flowchart, might appear in one of two formats: CPM, or critical path method, and PERT, or program evaluation and review technique.

Budgeting

Budgeting, sometimes resource planning, sets *resources vs time.* The resources we generally plan include

- Money—dollars
- Time—work-hours, work-months
- Material used in the deliverables
- Facilities and equipment used to produce the deliverables.

As an example, the strip of steel used in the punch press to produce a washer is the material; the punch and die set and the punch press itself are equipment.

In scheduling and in budgeting or resource planning, we must know how to estimate. Table 1-2 lists various techniques that can be used for estimating. It is important that managers in the engineering field learn how to estimate well, and how to define the project well. These two areas are not well done in many organizations. When discussing planning, we'll talk about the differences among project planning, personal planning, and group planning. Although we might *schedule* (i.e., *task vs time*) an individual project, the manager of a small group typically has a number of different projects under his or her supervision. It follows that the manager must combine the several projects into a *budget* (i.e., *resource vs time*) plan. Without considering all the projects together, the manager won't realize that, for example, each project leader has assumed that an individual is available to his or her project as that person's first priority, thereby totally overloading that one individual. Only by considering a composite of the various projects under your control can you ensure that your resources are best distributed.

ORGANIZING

The second of our functions, organizing, will be discussed from two angles.

Why Organize?

Why do we have to organize in the first place? Why can't we simply call in the people we want to work on the project, give them

TABLE 1-2.
Estimating Techniques for Engineering Tasks

1. Past performance—Historical cost records.
2. Past performance—Memories of what it cost.
3. Analogues—This is similar to that, and I have costs on that.
4. Supervisor estimates based on how he or she would do the task, then multiplies by a factor to adjust for the type of worker who will actually be assigned the task.
5. Supervisor asks someone else to estimate the job, then multiplies by a factor to adjust for the historical accuracy of that person's estimates.
6. Consensus method and other adaptations of management science techniques.
7. Negotiation with those who will be assigned the task; generally scope, specifications, and costs are negotiated simultaneously.

the specifications and other documents, and say, "Here. Go with it. Come back in six months when you're all done."

This would work well in a perfect world. Some folks do operate in self-directed work teams, but we still don't leave them completely alone for six months. To understand why we have to organize, let us look at Figure 1-1. On the left are your goals, the work to be done. On the right are your resources, your staff. At any one time, they are approximately fixed. The only purpose of organizational structure—and notice on the diagram we show them as elements in some type of network—is to match properly the resources you're "stuck with" with the work you're "stuck with."

Organizational structure is a network to match the characteristics of the work with the characteristics of the staff. For example, if you had one senior, two intermediates, and three junior engineers, we might organize one way: perhaps the senior and each intermediate would be paired with a junior, or the manager and the senior each might have an intermediate and one or two juniors working with them. But if you had six junior engineers, we'd have to organize a different way: here the manager has few options—either supervise all six juniors or try to pair those more proficient with those less proficient. A crash project with punitive or liquidated damages will be organized differently from a more routine project in which we control the schedule and the work is internal. So we organize for this one purpose: to make the project achieve its goals. This is the only purpose of organizational structure, and therefore, if you're asked what type of organizational structure you use, the best answer is, "Whatever's necessary to make it happen."

Span of Control

The second part of our overview of organizing is the question of how many people a manager can effectively supervise: How

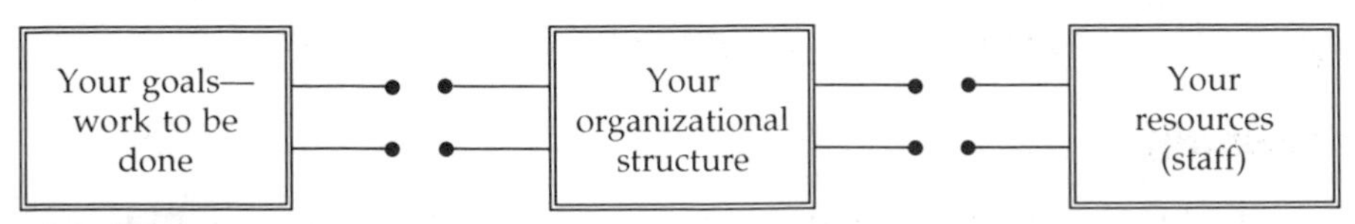

FIGURE 1-1. Why Organize?

does a manager compartmentalize the work and the people? What is the appropriate ratio of subordinates to manager? This is best understood through the principle of *span of control:* the number of people a manager can effectively supervise is a function of the nature or characteristics of that manager, the workers, the tasks, and the organization.

What is it about the manager that helps determine how many people he or she can effectively supervise? The manager's education, experience, and competency. What is it about the workers? The same three things, plus how self-starting they are. The more self-starting the workers, the less the manager has to supervise them, and the more subordinates that can be supervised by the same manager.

What about the tasks affects how many people can be organized under one supervisor? Is the task revolutionary or evolutionary? Routine or non-routine? What is the nature of the schedule, crash or more normal? Is the overall workload comprised of a small number of large projects or a large number of small projects?

The final aspect of span of control, the organization, includes such factors as sufficient support services, a not unreasonable amount of red tape, and your boss's management style.

Consider extreme examples. How many people could be supervised by a not very competent supervisor if they are not very competent workers, each having the wrong education and the wrong experience, on difficult tasks they've never done before, on a crash schedule, in a company that does not provide sufficient support? One. Tops. On the other hand, we've seen many cases where ten to twelve can be directly supervised by one manager. This is a big difference.

STAFFING

The third function of management is staffing, often called "human resources." Four elements make up this function: training, performance appraisal, job assignment, and recruiting.

Training

Training and development are among the most important roles of the manager. Optimizing the resources assigned to perform the work to be done is heavily dependent on continually optimizing the capabilities of the individual and the teamwork of the group. One problem that many managers have in training and development stems from the level of confidence we have in our staff. Since many managers were promoted from within the group they now manage, they know the people so well that they may have a tendency to emphasize their bad points and not give enough credit for their good points.

In addition, a first level manager is often promoted because (among other things) he or she was the best technical person in the group. They see some staff members as technically inferior and try to do too much themselves rather than spend more time on training.

Performance Appraisal

Elsewhere in the book (Chapter 13) is an entire chapter on performance appraisal; there we describe the formal appraisal—that once or twice yearly administrative system in which we prepare forms, hold interviews, get approvals, and so forth. But at this point, it's appropriate to mention what good managers know: in addition to well done formal appraisals, they must provide continuing feedback to their staff on how they're doing and encourage them to tell you how you can help them do better. Don't wait for the annual appraisal to provide feedback; that would be like coming in with a sack of comments about how the person has been doing things wrong for the past year. Instead, discuss the incidents as they occur to provide your staff with the opportunity to improve performance on the current task, as well as prevent the same problems on future tasks.

Informal performance appraisal is at least as important, and perhaps more, than the formal system because the informal system occurs all year. It gives people an opportunity to make changes as they go. In computer terms, the formal system is "batch processed," whereas the informal system is "real time, on-line." Informal appraisal is oriented more to events than to time periods. That is, we might speak to a person at the conclusion of a design review, at the conclusion of a phase of a project, at the conclusion of an important review meeting, rather than at the end of a preset period of time, as is the case with a formal appraisal.

Job Assignment

Job assignment revolves around one question: whether to develop people as specialists, assigning similar work to them again and again as they become more and more proficient, or to rotate people through the different types of work in your group. The more specialized a person is, the more knowledgeable he or she becomes, and the faster and better he or she can do a task. On the other hand, first of all, half the people don't want to keep doing the same things over and over again. They want a more varied work experience. Secondly, work doesn't come to us in neat, 40-hour-per-week-per-specialty packages. Sometimes one person is totally overloaded while another is nearly idle; that's the reality. Thirdly, when your only specialist in an area leaves or retires, it takes nine to eighteen months for the working group to recover. The fewer workers you have who can do a particular job, the greater your risk. And fourth, as the manager, specialization gives you less flexibility with your resources: each specialist is a fixed element. Lastly, the fifth point, staff have fewer opportunities for cross-fertilization of ideas because each works alone in his or her own narrow area. Such workers can no longer look at things in a systemic way, but only from their own narrow, component-level orientation: micro rather than macro.

One way of minimizing these problems is to make sure everyone has not only a primary area of expertise, but also a secondary or a back-up area of expertise. So every third, fourth, or fifth assignment, Specialist A works in B's area, and another time, B works in A's area. This minimizes the day-after-day boredom and the gap when someone leaves, as well as allowing the manager greater flexibility in allocating resources. We recommend strongly that in every area you establish this type of cross-training.

One other thing. You may need to sell this idea to some people who view it as a threat to their indispensability. Point out to them that cross-training will enhance their job security: if that special area is sold to another company, or if your company goes out of that market, or technology renders it obsolete, or business drops in that area, they still have skills to offer.

Recruiting

The amount of time spent recruiting depends a lot on general economic conditions. In a growing economy, you spend more time recruiting than in a time of reductions in force. Recruiting consists of a series of things necessary to acquire new staff:

1. Prepare job specifications and the necessary company paperwork including approvals.
2. Discuss with the Personnel Department the job requirements and any uniquenesses relating to the job.
3. Assist Personnel in developing recruitment strategies, such as in which journals to advertise, what key words to include in the ad.
4. When the résumés start arriving, review them and select the best; categorize them, for example, by salary desired, by years of experience, and so forth.
5. Conduct telephone screening of the best applicants to pare down the number that will eventually be interviewed.
6. Develop an interviewing plan with questions and topics and a weighted matrix for evaluation. By that we mean the topic *vs* our evaluation of how the person does in that topic.
7. Conduct the interviews, or if you're not doing it all yourself, arrange for interviewers. At this point, remember that if you're not the only one doing the interviewing, or if you're going to interview numerous people over a period of time, it's important that the interviewing pattern developed in point 6 be clear enough so that you can do a comparative analysis after all the interviews are completed.
8. Review the interviewing results.
9. Conduct follow-up interviews if required.
10. Assist Personnel in reference checks.
11. Assist in final selection of a candidate to be offered the position.
12. Assist Personnel in making the offer and in any negotiations; be available to answer the applicant's questions.
13. Develop an orientation plan for the successful applicant when he or she arrives.
14. Implement the orientation plan when the successful applicant joins your staff.

that. Otherwise input will languish in their in-baskets and we'll lose the time we gained. If the deviation is *zero*—right on plan—give the staff a "Good work!" and mention the next time we get together to review the project.

If the deviation is *negative*—behind plan—we have several things to worry about. First, find out why. Second, look for corrective action to get back on plan. If such action is possible, take it and monitor it to make sure it has the desired results. Modify, if required. But if no such action is possible, the worst thing is to kid yourself into thinking you're sticking to the now-obsolete plan. Your workers won't be sticking to it because there's no plan to stick to; everybody will be marching to different drummers, and you'll have organizational anarchy. If you can't meet the schedule, you have to bite the bullet, negotiate with your internal or external client, negotiate with your suppliers, negotiate with your staff, and figure out the best possible way to get through the situation. It won't be the original way, but perhaps it'll be palatable to the end user. It might even offer benefits of its own.

■ ■ ■ ■ ■

This concludes our discussion of the basic functions of the manager. Remember that these are the *functions* of the manager, but you have a big toolbox of techniques and methods to use for diagnosis and prescription. Innovation, motivation, communication, and performance appraisal are examples of these tools.

CHAPTER 2

DELEGATION

Three keys to success in management are training, motivation, and delegation. When you train a person, you increase that person's potential capability. When you delegate a task to a person, you give the person an opportunity to develop and use that previously latent, unused capability. When you motivate a person to do the best job possible, you're really optimizing the person's capability.

A technical analogy to not delegating would be an appliance not turned on. There is potential energy at the outlet, but no current is flowing. When you turn on the appliance and current is flowing, that's delegation. To make sure the appliance is operating efficiently, we may make a power phase correction. That is analogous to motivating the person to want to use his or her maximum potential in the execution of the delegated task.

WHY MANAGERS DON'T DELEGATE

Figure 2-1 lists the ten most common reasons managers don't delegate. They are discussed below.

"They will outshine me..."
"It'll take too long to explain."
"I like to do this work."
"They don't like to do this work."
"I'm the only one who knows how."
Control.
"They'll screw it up."
Fear of technical obsolescence.
"I can do it better and faster."
"No one else is available."

FIGURE 2-1.
Why Managers Have a Problem with Delegation

"They will outshine me . . .

. . . and my place with the company will be threatened if they do a great job." In fact, any time anyone who works for you does a great job on the task you delegate, your management ability increases in the eyes of your boss and your colleagues. Never worry about your staff outshining you. All it means (as far as your boss is concerned) is that you have a good group, you're managing them well, and you're getting productive output. Conversely, if they do a poor job (as far as your boss is concerned), you've done a poor job of managing the group. In your boss's eyes, your staff's performance and yours are directly linked.

"It'll take too long to explain."

Everyone has been guilty of this. You may think to yourself, "I should delegate this task to Sam, but he's never done it before. I could do this task in two days. I'll bet it'll take me two days to show him what to do, and it'll take him five days to do it. Then I'll have to check his work. That's poor economics." The facts as stated are absolutely true. It will take Sam a lot longer than you to do it, and it will take time for you to show him what to do and to check his work. But remember that everyone starts on a *learning curve,* the time it takes to come up to speed at a new task. From a practical viewpoint, if you're not willing to make such an investment in an employee, that person sees this job as a dead end. Others in the group are bound to see it the same way, and they'll want to transfer out.

This situation has an analogy in manufacturing. The time it takes for you to train someone, the time it takes for that person to come up to speed, and the time required to meet cost goals are analogous to one-time tooling and set-up costs. In other words, we have to be willing to take a short-term loss to make a long-term gain. Without it, your staff can't upgrade and your organization can't grow. The question to ask yourself is, "When's the appropriate time for me to take this short-term loss?" It should not be on a critical path task, for example, or on a task needed very soon. But certainly at other times and on other tasks, the risks are worth it and the long-term gains are absolutely necessary.

"I like to do this work."

This is a rationalization that most won't readily admit. In other words, some of you enjoy some technical tasks so much that you keep them for yourself. You are willing, however, to let your staff do the less interesting, more boring, more routine, "hack" work. You're paying for this pleasure with several big risks. First, your staff will realize that you're hogging all the interesting work. Second, you're creating a situation where no one else gets the experience of more challenging work; later on when something far more important comes along that needs your attention, you're stuck doing this other work because no one else knows how. So share the interesting work. You do it sometimes, they do it sometimes. Not only is your reputation as a manager improved by providing greater opportunities for your staff, but you will find it easier to delegate to them in the future. Lastly, sharing all the work facilitates synergistic cross-fertilization: each of you adds your own imprint and your improvements to the task. The input of many minds results in a geometric improvement pattern (something to look for in any work); the best you can do on your own is arithmetic.

"They don't like to do this work."

You try assigning a task to someone who says, "Come on. This is trivial, and you're insulting me by asking me to do it." So you back off. "Okay, okay. I'll do it myself or have someone else do it." How do we assign work that truly is dull and boring? No matter how state-of-the-art or how esoteric, every field requires some work that is truly dull, boring, and routine. There are several ways of delegating this kind of work.

☞ The most popular way that you should never use: give the dull, boring work to the low person on the totem pole, the newest person or the most junior person in the group. You may try to disguise it—"It is important for you to learn how to do this. It'll help you learn about what we do here." This won't work for long. For the first day or two, the person is interested. For the next year and a half, the person is totally unhappy and totally bored. Many of you have had this experience, so you know it's not good. Several other methods are much better.

☞ The second method (but the first of the good methods) is to find someone who enjoys this type of work. What is dull and boring to one person may be the spice of life for someone else. For example, one of the most common things that people don't like to do is writing reports. Engineers are notorious for trying to avoid writing reports and documenting things. But some people enjoy it. Some even love it. So try to find someone who likes to do the type of work you want to delegate. For example, hire a writer; provide someone to draft, rewrite, or edit reports out of your group—rather than the hard-pressed manager, who may not be a particularly good or comfortable writer.

☞ The third method is to share these tasks. Everyone will know that doing hack work is the dues they pay to get the interesting work. With everyone sharing equally in it, no one will feel put upon.

☞ The fourth method is to bundle all the tasks on the job together. Don't separate the interesting ones from the uninteresting ones. Give a person responsibility for a total job—all types of tasks, interesting and boring together. This has the added advantage of helping your employee feel empowered with the responsibility and authority for a total package.

"I'm the only one who knows how."

This is very dangerous! First, there's no one in your group with whom you can discuss these tasks, no opportunity for feedback.

Second, you're stuck with it, you have no options on assignment. Third, it appears to your employees that you feel they're incapable of learning the task. This is very bad for morale, but if it's true, you have some intensive training, some shifting around with other groups, or some hiring ahead of you.

Not to mention, heaven forbid, you should drop off the face of the earth before tomorrow! No one person, whether you or a member of your staff, should be the only one with certain knowledge; with promotion, transfer, resignation, illness, or death, that knowledge is gone.

Obviously, some special situations and unique tasks you'll have to handle. Almost all first-line managers, project engineers, and project leaders do a lot of technical work themselves. There's nothing wrong with you being the person who fills the void on projects, but you shouldn't be the only one who can. In fact, it is recommended that managers or multiple projects or more people generally *not* assign themselves to specific project tasks. You can use your time better for emergency tasks, to fill voids, to cover vacations, to address problem areas, or to handle marketing and other unpredictable tasks. Reserve your time for the unscheduled surprises.

Control

This is a big problem, particularly for younger managers. How much do you have to become involved in the details of the project to manage it properly? This should be addressed from two viewpoints: the task and the person receiving the delegation.

If the task is one that has never been done before, one that has been done before but has always caused problems, one that is on the critical path, one that is irreversible, or one that may be very costly, then you need to have control points and reviews closer together and in greater detail. Conversely, for more routine, simple, not very costly tasks or for those far off the critical path, the control points can be further apart and less detailed.

If the person receiving the delegation has never done the task before, is not your best person, is not a very good communicator, and is reticent about telling you the whole truth (or shades it), the control points should be closer together and in greater detail. Conversely, if the person is an excellent practitioner with a great performance record, regularly completes tasks, tells you about prob-

"No one else is available."

This is a problem, not a fear. You're a person who says, "I want to delegate, I should delegate, but I have no one to delegate to." This brings up the question of what to do if you don't have enough people.

Figure 2-2 examines the possibility of delegating to resources other than your full-time permanent staff. The top line represents the labor content of your anticipated workload. The shaded area represents the labor availability of your permanent staff. The gap between them is what you'll have to fill by external delegation—or

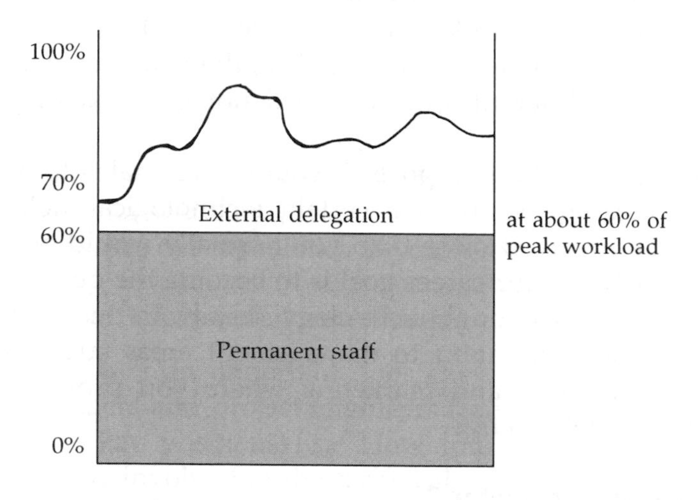

Above and beyond your permanent staff's normal workweek:

- Overtime
- Temporary, contract, job shop personnel
- Your own retirees
- Subcontracting
  - Engineering departments of subcontractors (e.g., equipment suppliers, component suppliers, etc.)
  - Architectural and engineering firms
  - Engineering consulting firms
  - Freelance consultants
  - Other parts of your company
  - Professors, graduate students
  - Moonlighters
- Co-op students, interns
- Part-timers
- Summer interns

FIGURE 2-2. External Delegation

slipping schedules, or changing the scope of tasks, or adjusting in some other way.

When it comes to working your staff overtime, remember that too much overtime can be counterproductive. To use overtime properly, make sure you schedule regular breaks; for example, if you must work 50 hours a week, it is best to do that for two weeks and then take one week off, do it for another two weeks and then take another week off. During your "week off," you work only 40 hours to give yourself a chance to rebuild your energy.

DEGREES OF DELEGATION

Figure 2-3 describes the degrees of delegation as follows:

At maximum delegation:
Take action, no further contact.
(You have delegated as much as possible.)

At the 2nd level:
Take action and tell me what happened.
(You just want to keep informed.)

At the 3rd level:
Look at the problem, tell me what you want to do, but I have *veto* power.

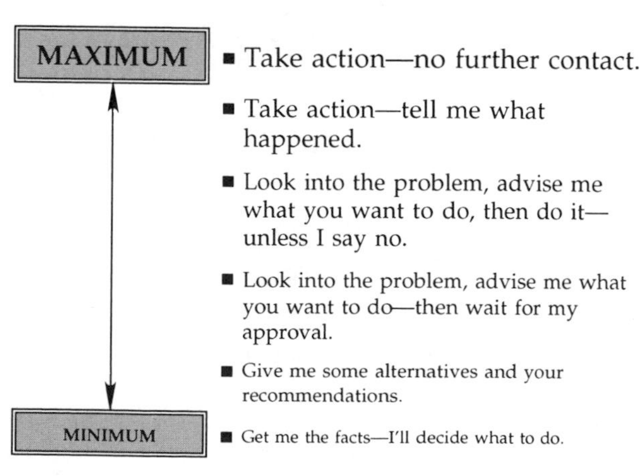

FIGURE 2-3. Degrees of Delegation

At the 4th level:

Look at the problem, tell me what you want to do, but I have *approval* power.

At the 5th level:

Give me some alternatives and your recommendations, and I will make the decision.

At minimum delegation:

You go out and gather the information for me.

CHAPTER 3
COMMUNICATION

It is said we are living in the beginning of the information age. But what good is information if it can't be used? To be used, information must be shared. To be shared, it must be moved. Moving information is communicating. Maybe we should say we are living in the communication age.

Good communication is a prerequisite for success in all engineering, scientific, and managerial functions. The process of good communication can be established by following a few basic steps. In addition, some common conventions, techniques, and discipline-specific practices have developed for special situations.

THE COMMUNICATION PROCESS

Communication is a five-step process. ☞ The first step is to *get and maintain the attention* of the entity with which you wish to communicate. In other words, do whatever is necessary to get it, her, or him to receive the message you want to deliver. It can be difficult to establish initial contact, more difficult to get undivided attention, and almost impossible not to lose some of that attention during the next five minutes. These are facts of life. Fortunately, ways of overcoming these problems have also been developed, and they will be discussed later.

☞ The second step is to *encode the message* in the form best suited to the situation. This is a mental process made easier by practice; it involves choosing the right symbols (language, words, structure, etc.) to express the message. Over time, as we become accustomed to familiar situations, common messages, and the same receivers, the encoding process requires less mental effort and becomes more automatic. This may seem to be good, but if we are not careful, we can forget how to formulate a message in a universally accepted format understandable by anyone. "Shop talk," "verbal shorthand," and the excessive use of abbreviations, acronyms, and technical terms are communication crutches we all use to minimize our mental effort.

☞ The third step is to *convey or deliver the message* using the most appropriate medium. Every message doesn't have to be written. Timely verbal messages are frequently more appropriate. Messages to robots and machines have to be in the electronic language they understand—speaking louder doesn't help with them any more than it does with people who don't speak your language. You may decide to use more than one medium of communication. For example, when delivering a message to a diverse group of people, you will increase reception if you project your message in both verbal and visual (graphic) form.

☞ The fourth step is for the receiver to *decode (translate) and understand the message.* This is another mental process accomplished by comparing the message received with information already known from previous experience, then determining its disposition. The receiver, for example, may decide the message is either false or irrelevant and ignore it. On the other hand, the decision may be that the message is important, valid, or useful and should be stored for future recall. This illustrates selective memory; we use or remember only a small portion of the messages our sensors receive.

☞ The fifth and most important step is to *establish feedback.* All forms of feedback accomplish the same purpose—verifying that the proper message was received. Good feedback must be bi-directional, involving both the receiver and the transmitter. This step "closes the loop" so that all involved know that communication has been accomplished and other activities can proceed. The communication process just described is summarized in Figure 3-1.

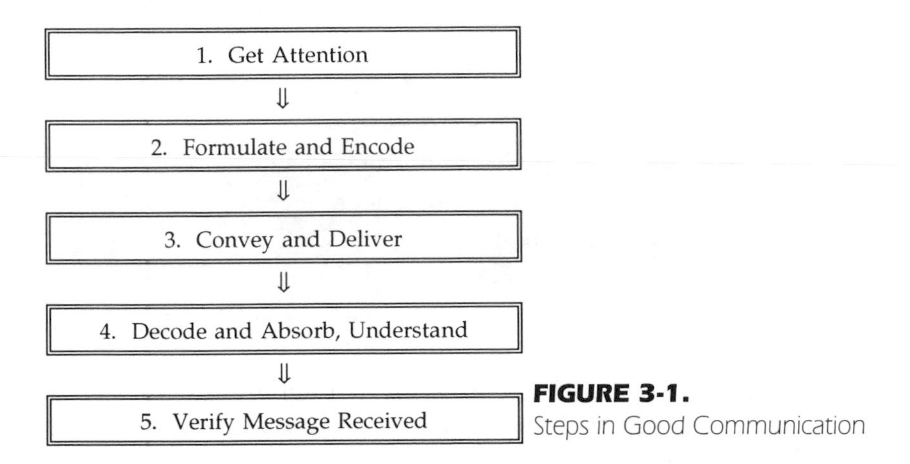

FIGURE 3-1.
Steps in Good Communication

COMMUNICATION BARRIERS AND OBSTACLES

We face many barriers and obstacles to good communication. The engineering manager today must deal with many people for whom English is a second language, work with people in different disciplines, supervise professionals whose education has been slim in communication skills, and must "translate" engineering information for non-engineering management staff. The ability to communicate may differentiate between otherwise equal candidates for promotion.

Memory—or the Lack of It. Communicating, following the five-step process described above, is difficult and has a low probability of success. If your brain's memory is your principal tool in the fourth step, you may accomplish only 5 percent of your communication objective. Not good or bad, this is simply the way it is. It is supported by research on how the mind works and experiments on what people remember (see Figure 3-2).

What is important is how to increase the probability of successful communication. The most effective way is to support your

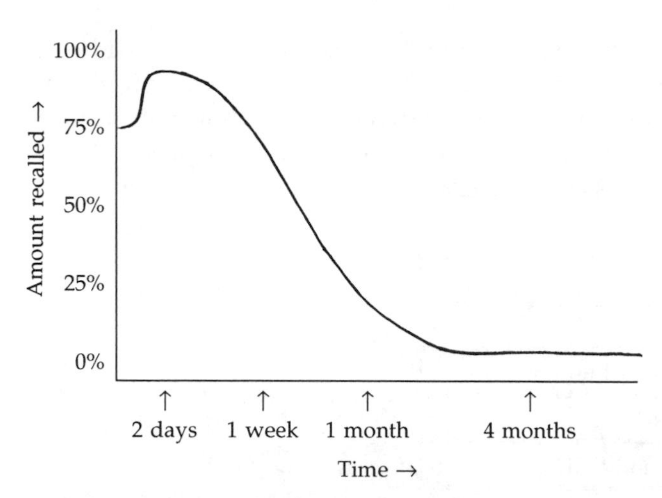

FIGURE 3-2. Memory Over Time

memory with notes, drawings, files, or a database so that your mind need function only at higher levels (i.e., conceptualizing and indexing) to recall details.

Mismatched Processing Speeds. Steps in the communication process proceed at different speeds. Mental processes are faster than speaking or typing (transmitting) and hearing or reading (receiving), and the electronic speed of some media can make conveying the message the fastest process step. What happens, therefore, is that communication proceeds at the speed of the slowest process step. Information traveling faster than this might not be received.

Dual Processing: A Communication Hazard. The human receiver is prone to a particular communication breakdown. Listening to (receiving) the message does not require all one's mental capacity, so the mind can (and frequently does) start working on other matters. In this "dual processing" the communication process may not have top priority.

Screening Information Through Expectations. In addition to the barrier of dual processing, the minds of both the transmitter and the receiver can erect mental barriers or screens to communication. For example, a person may have a preconceived opinion that the message will not be important, so the mind erects a screen of "not important" against the message. As shown in Figure 3-3, the message either cannot penetrate this screen, or penetrates it (i.e., is heard) only with a strong tendency to interpretation as unimportant. These screens are frequently based on bigotry, halo effects (explained below), stereotyping, and prejudice.

The Horrible Halo Hang-Up. This heading provides an example of how the halo effect can hinder communication. Those who like repetition in the first letters of words in a phrase (alliteration) will be "grabbed" by the title and want to read this paragraph. Those who are turned off by "cutesy" things will be tempted to skip to a more interesting section. This is the halo effect in action.

The halo effect is the name given to the influence exerted on the decoding process by prior experience or judgment. If you like something, it becomes important and you evaluate the message favorably and accurately. If you don't like it, it may not even penetrate your mental screen. In other words, the message that gets through to the receiver is only what he or she expects or wants to hear. A closed, unobjective mind is subject to the halo effect.

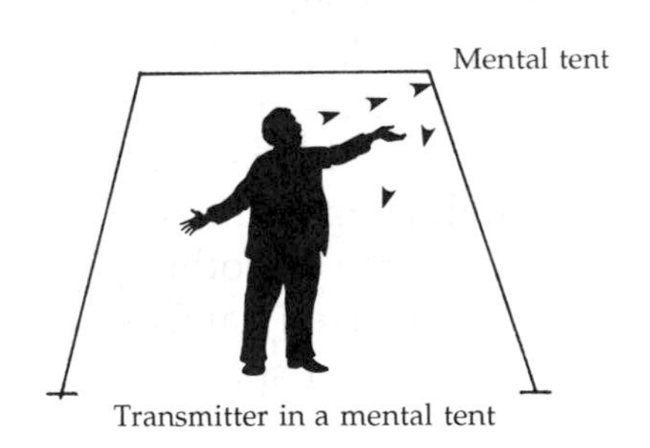

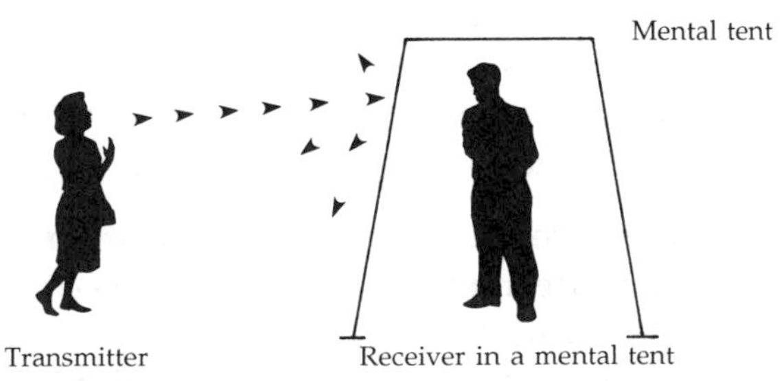

FIGURE 3-3. Screening Information Through Expectations

Receiver's State of Mind. Sometimes meaningful communication is not possible. When your receiver is emotionally upset or strongly preoccupied or depressed, unless you have remarkably good news, save it.

Be aware of the existence of these barriers, and take every opportunity to knock them down for instant improvement in your communication.

- Augment your memory and allow your brain to function at higher levels.
- Communication proceeds at the speed of its slowest component; information traveling faster than this might not be received.
- You're doing two things at once, and listening is your second priority.
- You hear only what you expect to hear.
- Halo effect—You perceive accurately, or take in at all, only what you think you'll like.
- You ignore information that doesn't fit your schema for that subject.
- Technical terms, argot, jargon, acronyms communicate only to a select few.
- Noise.
- Be aware of how past experiences and company legends might influence your interpretation of messages.
- Your English as a native speaker will need help with even very competent English as a second language.
- Blend careful listening to their words with careful attention to their body language.
- Be sure your receiver can listen.

TECHNIQUES FOR IMPROVING COMMUNICATION

Removing barriers is an obvious way to improve communication. It will get you out of the negative and up to zero on the communication scale. Proactive techniques such as the following will get you into the positive range.

The RAFT Technique. *RAFT—remember and fine-tune*—aids the communicating process by utilizing both sides of the brain to reduce the mental effort required to communicate.

The message transmitter need only remember the previously established form of communication. This can be done easily by calling up an image of the last communication. Let's say it involved a PERT (program evaluation and review technique) diagram. Knowing this makes it easier to fine-tune an encoded message using terms specific to PERT, such as "How much *float* is there for the task(s) just before *node 26?*" Several paragraphs would be required to ask the same question if tasks of the project were not shown on a PERT diagram.

The receiver also is aided by the RAFT technique. Referencing the PERT diagram either mentally (right brain visualization) or on hard copy or computer screen provides a fast and easy memory aid. Fine-tuning onto node 26 provides access to the detailed information needed to provide the answer (feedback) and complete the communication process.

Use Rich Media. A rich medium is one that allows simultaneous use of multiple information cues and symbols. The message can be reinforced, redundant, and simplified all at the same time. A face-to-face conversation, for example, is a rich medium. The message can be reinforced by body language, presented both verbally and visually (redundant), and simplified using graphics to display trends over time frames, sequential actions, and comparisons. A multiperson project progress meeting with presentations and graphic handouts is another example of using a rich medium to improve communication.

Make Communication Interactive. An interactive environment supports good communication in several ways. First, it encourages feedback without which the communication process is not complete. Secondly, it facilitates empathy (understanding and thinking like the other party), proper pacing, and "chunking." These are all necessary to accomplish good encoding.

A telephone conversation is an example of interactive communication. If the parties do not know each other, the first segment of the conversation usually consists of probing questions aimed at establishing a mental image of the other party. (And yes, stereotyping can be a problem.) Courtesy requires they agree that this is a good time to talk. Trial and error establishes the appropriate jargon and the pace most comfortable for all participants.

Practice Chunking. Chunking is the process of packaging information into small bits (chunks) for transmission and reception. The human mind, telephone systems, and many computer programs work with chunks. Telephone systems use "packet" (chunk) switches. Computers can be programmed in object-oriented languages in which the objects are chunks. Chunks are very common in our everyday world.

The antithesis of chunks is a linear stream of information. Unfortunately, linear streams are also common. Some teachers, professors, and lecturers present streams of facts, and engineers and scientists tend toward linear thinking. Older computer files were streams of data arranged in linear arrays. This is changing as rich, interactive media are introduced into classrooms and new technologies such as multimedia and hypertext become more common. The best communicators of the future will use chunking.

More Techniques for Improving Communication. Following is a summary checklist of techniques for improving communication. The best results, however, will come from selectively combining them until their use becomes second nature in your daily communication.

- Sensitivity to receiver
- Awareness of symbolic meanings
- Careful timing
- Feedback
- Face-to-face communication
- Many channels
- Reinforcing words with actions
- Simple language
- Redundancy.

HOW TO GET ALONG WITH YOUR BOSS

One of the most important aspects of communication is communicating well with your boss. The five conclusions in Figure 3-4 are based upon a study of some 100 engineering managers conducted by one of the authors of this volume.

☞ *Avoid annoying your boss.* Find out what causes your boss to be disproportionately annoyed, upset, or otherwise irritated with one of her subordinates. Ask yourself what causes you to be annoyed with some of the people who work for you, and you'll start to get a feeling of what might really cause your boss to become disproportionately upset. Once you know what these things are, avoid them like the plague! You may believe, intellectually, that this is petty nonsense, but if something has this negative an effect upon your boss, continuing could be counterproductive to your career.

Those 100 engineering managers voted this Top 10 list of things that cause them to really lose it:

10. Using the wrong communication techniques with the boss.
9. Indecisive—Can't make a decision.
8. Indifferent—Appears not to care.
7. Producing only if the boss constantly checks or pushes.
6. "End-runs"—Going around the boss.
5. Afraid to ask questions or say you don't know.
4. Being a "yes" person—Never being critical.
3. Being a poor team player.
2. Lack of honesty.
1. Being a perpetual pessimist.

☞ *Know and understand your boss.* Take the time to study your boss. Take the time to hear what other people say about him. Try some elementary "psychoanalysis" of your boss to help you under-

FIGURE 3-4.
How to Get Along with Your Boss

— Avoid annoying your boss. —
— Know and understand your boss. —
— Improve communication with your boss. —
— Timing is important. —
— *Honesty!!!* —

stand what makes him tick, the uniqueness of this one individual. The investment of your time will be repaid with a better relationship.

☞ *Improve communication. Timing is everything!* Consider these two points together. Ask two questions about your boss. First, when is the best time to see her? When is it easiest to get her (relatively) undivided attention? Second, when will you most likely get the answer you want? When is your boss the most amenable to your suggestions and proposals? After many years of studying this question, we have found that about 40 percent of bosses show a definite pattern.* In the course of a day, the pattern relates to when she first arrives in the morning, lunch, or after hours. In the course of a week, first thing Monday morning and late Friday afternoon seem to be critical. See if your boss shows such a pattern, and if so, use it to your advantage.

☞ *Honesty!* This is the last and most important point in communicating well and getting along with your boss. You must develop the reputation so your boss believes you are totally honest; you tell the whole truth and nothing but the truth; you will protect his interests; you'll be sure to bring to her attention anything she might omit; you'll attempt to be even-handed in your assessments—not overly pessimistic or overly optimistic; you'll make sure that if he doesn't have the chance to check with you, he doesn't have to worry that you'll act without his knowledge when you shouldn't, or delay critical action because he wasn't handy.

Ask yourself, readers, how would you feel about someone working for you whose honesty you questioned. You would feel compelled to over-control that person. Over-control takes time you don't have. To assure your boss of your scrupulous honesty, must you present her with every minor, trivial point? Obviously not. But anything significant or anything that you think might become significant must be brought to her attention. Must you take every decision to him? Again, no. Sometimes you'll tell him what you plan to do, not asking permission, not asking approval, but just to keep him informed. What bosses want is to be kept informed—no surprises.

---

*The other 60 percent showed no pattern at all.

THE BY-PASS

Figure 3-5 shows a straightforward type of organization. Level A is the second level manager, levels B and C are first level managers, and levels D and E are working people like engineers and technicians. The by-pass or end run is really an interrelationship of organization and communication. For example: Manager A is walking down the hall one day and runs into Engineer D.

Engineer D: Have you seen my boss, Manager C?

Manager A: No, but I know she's here; I saw her car in the lot. Why? Something important?

Engineer D: Very important. I just got a call from one of my suppliers, and if I don't get a decision phoned back by noon, the project might be delayed two weeks.

Manager A: Absolutely not. We can't accept a two-week delay in your project. It's vital to meet our goals for the quarter. Can't you make the decision yourself?

Engineer D: I normally would, but it's an $18,000 decision and I have only $10,000 signature authority. You know, if it were $12,000 or $14,000, I'd go right ahead and do it. But $18,000 is too much over.

Manager A: Well, I agree with you. But have you looked every where for Manager C?

Engineer D: I have, but I just can't locate her, and time's flying.

Manager A: Well, what do you think we should do?

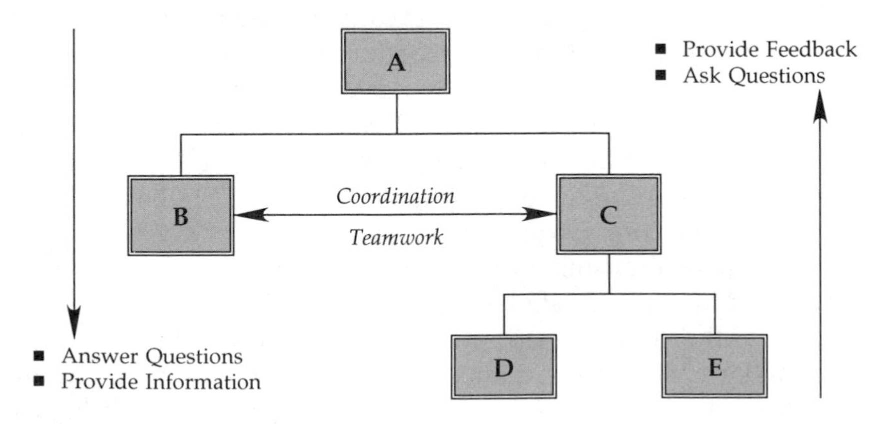

FIGURE 3-5. Organizational Communication

Engineer D: You are somewhat familiar with my project. Why don't we sit down, I'll tell you all about it, and perhaps you can approve my decision. That's the only way I can see of avoiding a two-week delay.

And so they did, and got off dead center. In and of itself this is not a problem, nor is the second example, when they again meet in the hall a few weeks later.

Manager A: Have you seen Manager C? I've been trying to find her all day.
Engineer D: She's here. We had coffee together earlier this morning. Anything I can do?
Manager A: As a matter of fact, there is. It's the last day of the month, and I've got to get my progress report to corporate. The only project I'm missing data on is yours.
Engineer D: I already gave my progress report to Manager C.
Manager A: Since I can't locate Manager C and I have to leave for the airport in about four hours, will you give me a copy and go over it with me?

Engineer D appreciated being able to repay the favor of a few weeks earlier and certainly didn't mind the interface with upper management. So again, they got off dead center. If such interactions happen infrequently, they are not necessarily bad things. But if they happen with greater and greater frequency and become much more the norm than the exception, Manager C is out of the communication loop. This is the by-pass or end run.

Let us examine why this occurs. Maybe Manager A used to be in Manager C's position, has been promoted up one level, but still maintains relationships with Engineers D and E. He hasn't symbolically cut the cord. Or perhaps Engineers D and E are trying to maintain the relationship with Manager A, and he doesn't prevent it. Maybe Manager C, who travels a lot, does not delegate her duties or authority when she's away; Engineers D and E may have to go to Manager A rather than leave something undone. Or Manager C gets so involved with her own pet projects that she's not as available to her staff as she should be. Perhaps Manager A has a warped and distorted method of control: he likes to check up on what Manager C tells him by going directly to C's staff, Engineers D and E.

Of all the things this author has observed in management, the by-pass is the most difficult to change. Very often it results simply from Manager A's impulsive nature. He thinks that the shortest distance between two points is a straight line, as it is: A to D is shorter than A to C to D. If you're in Manager C's position and often feel out of the communication loop, what do you do? The one thing you *can't* do is say to Manager A, "I forbid you to speak to D and E." A's response might be, "Why not, Former Manager C?" Nor can C say to D and E, "I forbid you to speak to A. Tell him to go away." They can't do that.

The by-pass is bad in two ways: you don't know what's going on, and it sort of deflates your ego. We can't do too much about your ego, but we can get you back in the communication loop. First, have D and E act as your sensors so at least you know what's going on. Arrange this by direct instruction to D and E, so they have no alternative: "When Manager A comes to you for information, assignments, interpretation, please leave a message to tell me what happened." You may be able to make an even stronger intervention: "If Manager A wants to assign work to you, remind him that I need to be involved; in fact, just send him to me."

One last thing: remember that this is a problem only when it happens a lot. Obviously there will be times when A going to D is most appropriate; it is equally appropriate for A to let C know about it. That isn't a by-pass, because you know what's going on.

RECRUITING

During times of growth the manager is often asked to spend a disproportionate amount of time recruiting new staff. The manager's ability to communicate well, particularly in interviewing, becomes critical. Interviewing should be called the fine art of listening, because it should be *no more* than 50 percent the interviewer speaking and *no less* than 50 percent the interviewee speaking.

Two types of interviewing are important to the manager, directive and nondirective. Directive interviewing is primarily used to obtain facts. For example: Did you ever work on high-frequency power supplies? Answer, Yes. How many years' experience have you had? Answer, 14 years.

One of the techniques helpful in the nondirective style is asking how a problem could be solved and having the person think out loud to the solution. Another is the open-ended question: Where do you see your career in five years? You can probe: Why do you want to work for this company? You can ask about a person's experiences: Tell me about what projects you have managed during the last two years, and the uniqueness of each one. Finally, you can ask a person's opinion: What do you think of this approach? Nondirective interviewing is used when you're looking for feelings, solutions, facts, or when you're trying to find out this person's viewpoint.

The task of hiring generally follows the sequence shown on page 9. Please remember that, at the same time the applicant is selling himself or herself to you and your organization, you are selling yourself (as boss) and your organization to the applicant. If you decide to offer the position, you want to be accepted.

All of this requires a good deal of communication ability. Pay particular attention to clarity, and pay particular attention that you do *not* make offers that are beyond your authority to make. Take care to describe the work environment accurately; don't foster unrealistic expectations so that your new employee comes to feel you misrepresented the job, work group, or organization.

4 CHAPTER

MOTIVATION AND INTERPERSONAL RELATIONSHIPS

LEADERSHIP AND MOTIVATION

To understand the connection between leadership qualities and motivation, let us use McGregor's famous Theory X—Theory Y. Theory X, the Authoritarian Manager, is modeled in Figure 4-1. Today we're more likely to call this person a dictatorial manager. We notice that she really believes that people don't want to work and will try to get out of doing it. If you want them to work, keep an eye on them. Turn your back and they'll stop working. People only work when they're forced to. People want minimal responsibility and maximum direction. People are no good. Best course of action: get them before they get you. These are the Theory X manager's beliefs.

Theory Y, the Participatory Manager, is more likely to be called a democratic manager today (Figure 4-2). He believes work is natural; whether you watch workers or not, you'll get the same amount of work. He believes that people have a strong need for

down. People may have transferred out and transferred in. All these will influence how the group can best be managed and what style to use.

We change with time. Ask anyone who's been a manager for a long time. Ask how she has changed since the beginning of her managerial career. Most people will say they've become more relaxed and mellow as time goes on. Not because they are no longer as interested, but because they now have far more self-confidence in their managerial ability, a better "toolbox" of techniques, and a higher level of competence.

Finally, the project or work situation affects your style. The most common agent of change is an emergency—a threatened cancellation of a major project, a flood, or a strike. What happens to the manager's style? In most cases, pressure tends to push us toward Theory X.

In conclusion, there is no universal management style. You must modify your style as these four factors change. Your ability to adapt as needed will markedly affect the motivational climate in your group and, thus, its performance.

MASLOW'S MOTIVATIONAL MODEL

Perhaps the most universally known motivational model is Abraham Maslow's hierarchy of motivational needs. It has been the basis of all motivational thinking in the industrial setting since it was first presented. Maslow was one of America's great post-World War II psychologists. In simple terms, his approach means that the key to motivation is to know what a person's needs are and satisfy those needs.

Knowing each employee's needs is a tremendous burden on managers. Certainly people don't come to us every morning and drop a little note saying, "Hi! Here's today's open needs. Please fulfill them." Management's role is to create an environment, an ambience, in which people are willing and desire to speak with you about things that are on their minds and things that are troubling them. This enables you to know some of their specific needs and to understand better how they work.

If you are an unempathetic, unsympathetic, uncaring, not-willing-to-give-people-time kind of supervisor, you're not going to know your subordinates' needs. Conversely, if you are empathetic, sympathetic, caring, willing to give people time, willing to listen carefully, and willing to be confidential, then you'll find out what's going on and you'll know the needs of your people.

Maslow stated that human motivational needs exist on a series of five steps or plateaus. Figure 4-3 displays them as steps in a pyramid, but you can also imagine them as stepping stones coming up to your front door.

☞ The lowest level of needs in this hierarchy are called *physiological or survival needs.* In the 1990s, this means very simply food, clothing, and shelter—as basic as you can imagine.

☞ The second level, *safety or security needs,* include protection services such as army, navy, police, fire fighters, hospitals, and others to provide medical care, flood control, clean air, and clean water. Another part of safety and security needs is job security. Levels 1 and 2 together constituted what Maslow called "basic" or "primary" needs.

Consider level 1 and 2 needs in a work environment. As managers of primarily professionals, technically and administratively skilled workers, very few of you are dealing with minimum wage employees, the "working poor." Therefore, we will assume that food, clothing, and shelter needs, at the minimum, are satisfied. There will be exceptions, severe health problems that drain finan-

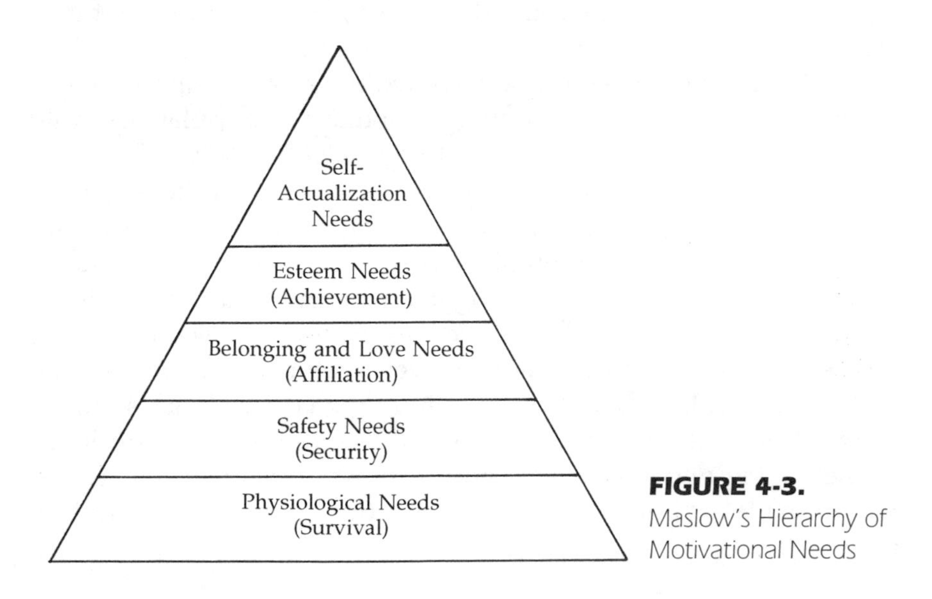

FIGURE 4-3. Maslow's Hierarchy of Motivational Needs

cial resources, child and spousal support among them. But for the overwhelming majority, their paychecks take care of the minimum.

Safety and security needs, via both government agencies and company benefits, are provided in most cases.

So what needs are left at levels 1 and 2? The only one you're likely to have to deal with as a supervisor in an industrial or governmental setting is job security. During some periods of time in any economy, in a given industry, or in a given company, job security is no problem. At other times it's a constant problem. For example, IBM went from roughly the end of World War I to 1992 without laying off anyone; during all that time, job security was never a factor at IBM. Eastman Kodak and other companies went for decades with no job security issues. Yet employees in these companies and across entire industries saw their job security evaporate in 1993 and 1994 when the Soviet bloc collapsed and the U.S. government subsequently realigned defense spending.

☞ The third level, *belonging and love needs* (affiliation needs), includes being part of a family, a circle of friends, and a greater community, and being in a profession, trade, or craft. The common denominator is being recognized as a part of a group.

☞ The fourth level, *esteem needs* (achievement needs), includes such things as ego, job satisfaction, job recognition, and feelings of pride, success, and accomplishment. The common denominator among these is I, me, the individual—not part of a group, but me, me, me, me, me.

☞ For the fifth level, Maslow coined the term *self-actualization.* It has become part of the working vocabulary of popular psychology. A term with greater meaning today is *self-fulfillment.* This person has achieved everything he or she wants out of life—materially, spiritually, physically, emotionally, across the board. In practice, how many such employees is a supervisor in a Western industrial or governmental setting likely to find? Very, very few.

Of level 1 and 2 needs, only job security is an issue of consequence to you as a supervisor, and level 5, self-actualization, rarely presents a problem except as a continuing goal for all of us. Aside from job security, what managers most commonly see are level 3's belonging and love needs and level 4's esteem needs. These will provide opportunities for motivating people in technical environments.

You may have to move either up or down in this model. Perhaps last year everything was going well and your group operated at level 4. On levels 1, 2, and 3, all your subordinates' needs were fulfilled. But today, business has taken a bad turn, your personal life has taken a bad turn, and you're coping with level 2 and 3 needs. It is a reversible model, meaning that once you go up, you may not stay up. And thankfully, once you go down, you may not stay down.

How does this work in practice? Table 4-2 presents five connections between Maslow's theory and our practice. These five points do not all carry the same weight. Factors 1 and 2 are 90 percent of it, so let us address those. A satisfied need no longer serves as a meaningful motivator; more is overkill. Or, if I never needed it in the first place, giving it to me will give you no meaningful advantage and no payback. And understand that higher order needs do not operate as meaningful motivators as long as lower order needs are not satisfied. In other words, you get the most motivational return when you satisfy someone's lowest level of unfulfilled needs.

TABLE 4-2.
Maslow—Theory to Practice

1. A satisfied need no longer serves as a meaningful motivator.
2. Higher order needs do not operate as meaningful motivators as long as lower order needs are not satisfied.
3. In industry, most people work on the job because of their physiological, safety, and social needs.
4. Ego needs are rarely satisfied adequately.
5. Very few of the jobs held even by upper echelon business people offer an opportunity to satisfy the need for self-fulfillment.

When Motivation Won't Work

At level 3, affiliation needs, all too often things are happening in someone's personal life that you are not aware of. And even if you were, there wouldn't be a darn thing you could do. Be aware of the kinds of personal events that may prevent you (for a period of time) from motivating this person to any degree.

An illness in the family is a good example. It might be terminal illness, hospitalization, or unidentified during the period of medical testing. Is this person's mind on the work? Of course not! Yours wouldn't be, either. With a serious medical problem, it may be months before that person's mind is back on work. You should not say—or even imply—"This is not a charity. You've got to produce." The person can't produce. If you are too brusque, too demanding, inhumane, other people will know about it and you'll be marked as "heartless." All you can do during that period is to support the person, help facilitate the resources available through the company's personnel and insurance functions, and be sure that you're not overtaxing him or her by requiring overtime or travel. This is particularly important with an employee who has proven to be loyal and a hard worker. You may have to change or modify the job assignment during the difficult period.

Will the other people in the group object to picking up some of the load? Generally not: "There, but for the grace of God, go I." If they were in the same situation, they would like the same consideration. However, this cannot go on forever. Even if there's a death, the bereavement cycle can only last for so long. If someone seems unduly affected or affected for an overly long period of time by a difficult life event, he or she can benefit from professional mental health assistance.

The second example is a person who is in the midst of a marital problem. It may be difficulty with a spouse, a question of separation, divorce, child custody, or a post-divorce spousal or child support problem. Any of these will affect a person's concentration. Again, you may have to steer your employee to the personnel department or other available resources. And this may go on for months. What you have to do is work around it, assist the person as appropriate to come out of this "nose dive," and make sure the effect on work diminishes as quickly as possible.

The last example has to do with a situation that you should never try to handle by yourself under any circumstances. It is when you suspect a person in your group may have a substance abuse problem. First of all, as technical managers, we don't have the expertise to decide if the person has or has not a substance abuse problem. All we can say is that a person's behavior or performance has changed. Changes in dress, punctuality, absenteeism, cooperativeness, work quantity and quality can be affected, and these are things you can comment on. Beyond that, what you have are no more than suspicions. In the case of alcohol abuse, an odor may be obvious. Still, for legal reasons, make sure that you involve someone in Personnel who's an expert in the area. Your employee may need assistance, but you need assistance in how to provide it. Get the best assistance available through your company or through community agencies.

HERZBERG'S MOTIVATIONAL MODEL

The motivational model that has been found most useful for both the typical, first line formal manager and the informal non-authority manager is Frederick Herzberg's "motivator-hygiene" model. Herzberg began with Maslow's premise: the way to motivate a person is to satisfy an unfulfilled need. Herzberg then developed a checklist of the factors that seem to cause motivation and demotivation.

Until this work, it was generally thought that any factor that caused motivation could also cause demotivation (see Figure 4-4). For example, if you receive little recognition for an outstanding job, you certainly are demotivated. On the other hand, if you receive a moderate amount of recognition, you feel you receive what is due to you, and therefore, you are neither motivated nor demotivated—sort of neutral. Alternately, if you receive lots of recognition for an outstanding job, you truly are motivated.

Herzberg's model posits two sets of factors to consider *vis à vis* motivation: motivators and demotivators. (Actually, he called the demotivators "dissatisfiers, sanitary factors, or hygienic factors.") Demotivators are like dirt gumming up the machine. Sometimes called a "two-factors approach," this model is the foundation for current thinking on motivation in the industrial setting.

Figure 4-5 lists the factors that cause motivation and demotivation. Bear in mind that these are Western models based upon studies of people in the Western industrial environment. Both the Maslow and Herzberg models carry cultural and sociological baggage. As we deal more and more in the global economy, these

The form of recognition depends on you, your group, your company, your environment, and so forth. You could simply say it. You could put in writing to the person. (Sending a copy to your boss seems to have a geometric effect.) You could give the person publicity and visibility by having him as a co-presenter with you at a staff meeting on the project's progress. You could give her the opportunity and funds to write a paper and travel to a meeting to present it. How about providing the clerical, writing, and graphic assistance to help prepare a paper for publication? You might be able to give extra time off or a job assignment your employee really wants. Remember that motivation is individual in nature; choose those things that will work for each person.

Advancement. Advancement motivates by offering job promotion and job growth. When the economy is particularly bad and organizations are getting smaller, this is not as available as it would be in a time of growth. In any case, the lower level supervisor does not do the promoting. In larger companies and in government, it's part of a big bureaucratic procedure. Even in small companies, it requires at least one signature besides the immediate supervisor's.

What is the role of lower management in assisting the job growth of our staff? First, identify a person in your mind who you think will be ready for promotion in the next year or two. Second, give the person any missing experience or education. You must be knowledgeable in the ways of your company to know what is required for the person to be seriously considered. Third, give the person lots of visibility and publicity so those who make these decisions are totally aware of his or her potential. And fourth, when the promotion opportunity arises, back the person. Be that person's sponsor, protagonist, and ambassador.

Responsibility, Authority, and Control. These three act together. As a unit, they are particularly significant for relatively newer and relatively younger employees. When a person joins the organization, you're understandably a little cautious, and you tend to give assignments that the person won't fail. You give more short-term assignments, you provide more control, and you don't hand over total authority at the beginning. As the employee develops and proves what he or she can do, you feel comfortable turning over more authority and exerting less control. Your employee will notice this: "I am quite happy that my boss's confidence in me is increasing. Therefore, I'm going to even do an even better job *[either]* to repay this confidence *[or]* to get him off my back." This approach is an indirect form of recognition.

Motivators revolve around the content, the core, and the internal workings of the job, whereas demotivators concern its context and surroundings. Motivators reflect how one feels about one's job; demotivators how one feels about one's company.

Who controls motivation? Who controls the work itself? Who controls how work is delegated within your group? You do! Sense of achievement—you set the stage for it. Recognition—all yours. Advancement—all yours. Responsibility, authority, and control—all yours. As the demotivators are within the control of upper management, the motivators are primarily within the control lower management.

Now that you know where you have control and where you don't, do not spend all your time beating your head against the wall, trying to remove the demotivators. All the demotivators can never be removed; economic considerations generally prevent that. Do spend some time getting upper management to minimize the demotivators, but spend far more time where you have the unilateral control to maximize the motivators. Strive to tip the balance in your employees' favor. If you can create a situation with far more motivators than demotivators, then the balance is very definitely in your favor. Most employees in your group will think, "I know there are problems. I wish there weren't, but there are so many advantages."

BURN-OUT

Any discussion of motivation would be incomplete without considering burn-out. Burn-out can take either of two forms. The first is about the job, the second is about life, and they often occur simultaneously.

☞ *Job burn-out* occurs when people have been on one job too long. To understand the burn-out phenomenon a bit better, refer to Figure 4-6. Here we see that all of us in the technical fields fall into one of three categories.

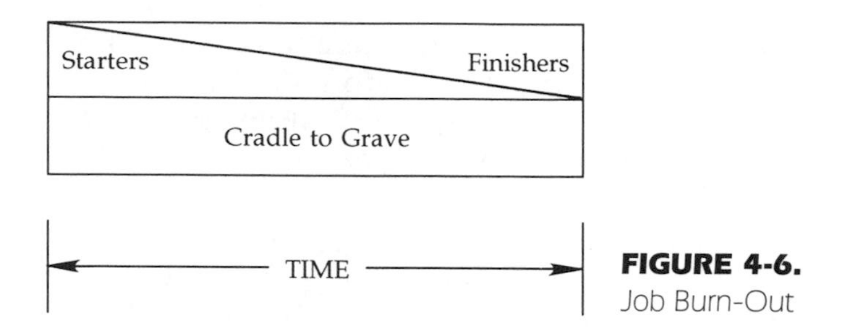

FIGURE 4-6.
Job Burn-Out

The *cradle to the grave* people are those who want to be on a project from beginning to end. No matter how long it is, they never lose their initiative, their drive, or their enthusiasm. They offer the continuity required on a project.

The *starters,* who usually are the largest group of technical people, love to be there when the paper is blank. They don't want anyone else's imprint on the design. They want to conceptualize and visualize a job, and let it follow their thinking. But when the job becomes routine, unfortunately, they seem to lose interest; their motivation may wane after as after as little as six months. They have no interest in routine things like drawings, reports, documentation, training, spare parts, maintenance, manufacturing engineering, and so forth. Starters are like the solid rocket boosters that provide the one-time thrust needed to achieve orbit; they are the push that gets a project going.

Finishers, the third group—and unfortunately the smallest group—are those people who are a little intimidated by a blank piece of paper. "Tell me what to do, and then I'll do it." But they know how to cross the t's and dot the i's. They're the people who worry about tooling, manufacturing engineering, documentation, reports, training, and all those important things necessary to complete a project properly. In fact, the lack of finishers is cited as one of the reasons that quality has been so low in the United States. Without enough finishers, we never really complete a project.

To staff a project properly, the manager must know about job burn-out. Any project must have enough starters, enough finishers, and enough cradle-to-gravers. Notice in Figure 4-6 that the number of starters slowly diminishes because some starters lose interest in a relatively short time; others last longer. It isn't that we take a starting team off and put a finishing team on. Team composition is in constant flux. Remember, if everyone were cradle-to-grave, we wouldn't need to discuss this at all.

☞ The second kind of burn-out prevents a sense of achievement from being a major source of motivation. It is known as the *mid-life/mid-career* crisis. Usually it happens to folks between ages 35 and 45 and comes with such thoughts as, "Is the second half of my career going to be like the first half?" And in the broader spectrum, "Is the second half of my life going to be like the first half?" Sometimes a person goes through a period of malaise, a lack of interest, during this period of introspection. The individual may not even be aware of its cause.

Occasionally, one hears about a person who (as an example) gives up being a systems analyst in a computer company in Silicon Valley for berry farming in Oregon or fishing on Puget Sound. People usually don't redefine their goals this radically, but changes in geographic location, spouse, or attitudes toward life are not uncommon. A mid-life/mid-career crisis is not something that the individual causes, nor does the company. It just happens—probably as a result of a complex weave of sociological, psychological, and maybe even biochemical forces we haven't the knowledge to unravel. As managers, all we can do is lend an ear if it's requested.

MOTIVATION AND AGE

The question of age as a consideration in motivation is important. There are two schools of thought. One says it's the formative years in people's lives, ages 10 to 20 or 15 to 25, that are significant, and how those years were spent will affect them for the rest of their lives. The more drastic the circumstances in which they grew up, the more extremely they will be affected. Accordingly, having grown up in Nazi-occupied Europe or in a region recovering from a severe earthquake might remain an important part of one's life.

The other school of thought says that the current period in a person's life is more important—one's twenties, thirties, forties, and so forth. In stereotypical terms: In the footloose and fancy-free twenties, you change jobs every two years and spend all your money on a sports car. In the thirties, marriage, family, and mortgage capture your attention. In the forties, the children are in college, expenses are even greater, and a mid-life crisis might seriously distract you. Except for that very much different body, the fifties are quite like the twenties: the mortgage is paid off, children have left home, and you're vested in the pension plan. Both are periods of less responsibility than the thirties and forties.

This second school of thought must be modified by the changing patterns in the population that *you* are managing. At least three factors are vastly different now than a generation ago. First, the increasing divorce rate results in more people having two families, perhaps spreading out their parenting over many more years. For example, a person marries, has two children, divorces when the children are 6 and 9, marries again, has another child, and now has children ages 1, 8, and 11—a fairly realistic example of its type. This person will be a parent of minor children for 28 years! 32, if you count the college years.

Second, many people decide to begin their families at a later age. If our employee from the previous paragraph had her first child at 31—increasingly common—she could have tuition payments until her retirement!

Finally, employment is less stable now than (arguably) since the Depression. Mergers, bankruptcies, closings, consolidations, downsizing, eliminating product lines, changing technologies, geopolitical factors ... most employees will have many more employers during their careers than their parents and grandparents did, and many will have more than one career. This employee we've been thinking about for the last few paragraphs, the one with three kids and tuition until retirement? She might be able to manage well if her employer downsizes her out of a job when her youngest is 3—and the oldest is 5 years away from tuition, and she is 44 and just hitting her professional stride. But what if her employer gets consolidated when her oldest is starting law school—and the middle one is needing tuition, and at 53, the human resources people keep thinking they ought to take that other phone call just as they reach for her résumé. What then?

Let these two schools of thought complement each other in your motivational analyses. Following formative-years thinking, pay particular attention to (a) those members of your staff who have grown up in a different culture where the formative years, for example, have a different meaning or different influences, and (b) those members of your staff who had some unique experiences in those early years. Following current-period thinking, it is incumbent upon you to understand the personal lives of the people working for you—within the limits of privacy. Everything considered, age certainly is a factor in how you motivate people.

MOTIVATION AND GOALS

Goals can be a source of motivation, provided they are not so difficult that they are considered impossible and not so easy that they offer no challenge. Research has shown that, when given a task so routine that it becomes totally boring, most simply do it and don't attempt to find a better way. When given a task that seems impossible (whether or not it is), many simply give up. But we grow only by being challenged. We must continuously challenge how we are doing things and look for ways to improve the process.

SPECIAL CASES OF SUPERVISION

The Superstar

Consider supervising a superstar or an authentic genius (Figure 4-7). At first you might say, "I wish we had more of them." Then you realize that, for example, although brilliant, this person is the world's worst at interpersonal relationships. How do we use this person in our people-oriented project work? It isn't easy.

In most cases, the superstar who has much to contribute—not the prima donna who only thinks he's a superstar, but the authen-

- The superstar
- The recent graduate
- Technical obsolescence
- The first
  [check all that apply: ☐ woman, ☐ man, ☐ Japanese, ☐ unmarried, ☐ African-American, ☐ Hindu, ☐ parent, ☐ wheelchair-bound, ☐ Kenyan, ☐ dwarf, ☐ Native American, ☐ dyslexic, ☐ Druid, etc.]
  in a group that previously had none
- Former colleagues
- Older subordinates
- Soon-to-retire subordinates
- Malcontents
- The "blob"

FIGURE 4-7. Special Cases in Supervision

can train that person; it may be too late for that. You may be better off looking for a place in the overall organization where that person's present level of knowledge, attitude, and reputation can succeed. Someone who may no longer be the best R&D person may understand the company well enough to work in some other area. Someone who may no longer be the best designer may have excellent contacts among suppliers or customers. Try to look at the needs of the overall organization as well as the current state of development of the individual.

Your Former Colleagues

Although it can be difficult, this is something we all hope we'll be doing several times in our careers. Suppose you've been promoted from within your group to supervise your former peers. Or perhaps you transferred to another group and are now returning to supervise a group in which you were once a member. This causes emotional problems for many managers. First of all, there are social relationships. Some of your new subordinates may be close friends; others you never liked. How do I appraise the performance of my buddies?

You also have certain advantages. By knowing the people better than a stranger would, you can better help them and the organization. You know what to say and what not to say because you don't have foot-in-mouth disease with these people. And you know exactly how to correct them without offending them. You know how *not* to deal with them.

You had lunch with Sam and Jane most every day for the last four years. That can't continue. You must be sure to have lunch with everyone at various times. The one thing you don't say to yourself is, "I can no longer socialize with my former colleagues." This would create a class and status gap, and your former colleagues might think you felt too good for them. Just be fair and equitable in your relationships. Take advantage of the knowledge you have, and most importantly, don't let them become your former friends. You need their friendships for success. You need to maintain a good level of friendship and comradery with them. When the chips are down and you're in a spot, you can call upon their help, not on the basis of "Do it for Company XYZ," but "Do it for me, your buddy, who needs your help."

Older Subordinates

Some older subordinates may feel that they themselves are failures because you, a much younger person, got the job. This is particularly true if you have the same education or background. One of the things the younger manager must be careful of is not to rub it in.

A younger supervisor would do well to ask advice of older subordinates, particularly in those areas where they may have more knowledge of past events in the company because they've been there longer. Treat this knowledge gained as a result of longer tenure just as you would treat any knowledge gained as a result of any other worthwhile characteristic.

Appreciate the fact that they are willing to share their experience and their knowledge with you. Don't imply that you're dependent on them, but rather that you're open to all suggestions and willing to listen to all considerations.

Soon-to-Retire Subordinates

This can be a great challenge even to the very experienced manager. A minority of soon-to-retired subordinates believe they have earned a vacation: "OJR"—on the job retirement. What can you do to motivate these people? How can you motivate people who believe you won't fire them if they don't put out enough work? They know all the legal aspects of firing a long-term, older employee who is close to retirement.

This challenge is similar to that posed by your older subordinates—to create a situation in which they feel they're contributing something unique and special. For example, they may do particularly well with training younger people because they have a feeling of responsibility toward the newer members of the staff or newer practitioners in the field. Perhaps they visualize their own children in a similar situation.

Look for a position in which the soon-to-retired subordinate's experience and historical knowledge is of great importance. Make sure that they never work by themselves, particularly those with a lot of knowledge to leave behind. They won't sit down with a tape recorder and begin, "Okay, here's everything I know." It's important that they work with other people whom they know respect their abilities and will carry on what they learn.

Malcontents

Be very careful of the malcontents, those who hate the company. In a team environment, they may attempt to poison their coworkers' attitudes: "This is a lousy company to work for. I've worked here 35 years, and they've been unfair to me every year." The malcontents require special handling. They are, happily, a minority.

The Blob

He breathes, and that's about all we can say. When you have an employee who isn't producing, you've got to take action to protect both yourself and the group's performance level. When that person produces below the group's minimum acceptable level, you have a problem. If you accept such low performance, the message to your hard workers is that their manager is weak, refuses to deal with this issue, and doesn't care that they have to pick up the slack.

Ask yourself, does the blob produce above the group's minimum acceptable level? You must know the difference performing below everyone else—someone has to be last—yet above the minimum acceptable level. Don't waste your time trying to convert or motivate someone whom you and other supervisors have failed to rehabilitate. It's not worth it. You are better off speaking to your manager about how to move this person out of the company. Too many managers spend too much time trying to rehabilitate where there's no hope of rehabilitation.

CHAPTER 5

LEADERSHIP: WHERE STYLE MEETS SUBSTANCE

Leadership is often spoken in the same breath with management. And when we discuss either of these, we often talk about style. Analyzing the style someone brings to the management and leadership process is a common way to understand their approach to and effectiveness in dealing with others. Style analysis has become a major component of management and leadership development programs.

The style analysis approach offers three main contributions to our understanding of management and leadership:

One's primary management and leadership style is a result of one's personality.

Styles can be learned and then adapted to fit the challenge of the moment.

There are distinctions between management and leadership: although often embodied in each other, they can be described differently.

STYLE AS A RESULT OF INDIVIDUAL ATTRIBUTES

The word "style" is used to describe the particular flavor someone brings to their activities. We say that people dress with style, act with style, or speak with style when they distinguish themselves with some unique look, behavior, or emphasis.

Leaders display style when they capitalize on certain traits or attributes so that others readily identify with those and respond with zeal or even passion. Martin Luther King, Jr., was a leader who engendered passion because he brought a charisma to his speechmaking that aroused, motivated, and excited his listeners. His "I Have A Dream" speech, delivered at Washington Monument Park in 1968, is still studied and revered as an example of brilliant oratory. In it he revealed his particular vision using a unique and stirring form of rhetoric.

Charismatic, inspirational behavior is one way leaders demonstrate their unique style and manage their approach to influencing others. There are other ways. Not all leaders, of course, have grand visions of universal appeal or access to national platforms. Some leaders respond to a calling more modest and more routine than changing a nation's mind about a social condition. Everyday business and organizational leaders, though, can learn something from the way world leaders use rhetoric, vision, behavior, and action to inspire others.

Individual leadership style may emerge in different ways depending on the personality traits of the particular manager or leader. Your style may be a natural and inevitable result of your personality as you interact with others, not related particularly to any conscious effort. This kind of leadership style is indigenous and natural. It may be displayed unconsciously. On the other hand, when we say that we wish to develop our indigenous management or leadership style, we mean that we want consciously to enhance characteristics and attributes to optimize our unique style.

A highly developed individual leadership style may result from optimizing personal attributes and characteristics that are perceived by others as irresistible. Thus, Reverend King may have recognized (i.e., made conscious) his ability to appeal to others through rhetoric and oratory, then consciously developed that skill. This is something preachers are taught to do.

Some managers and leaders study and develop their style and use it all the time. They become walking caricatures of themselves; their particular style is their logo, their banner, and they use it all the time in essentially the same way. U.S. presidents do this now, as instructed by their media consultants, to keep their image before the public as consistent and predictable as possible. When in a position of such visibility, changing one's style in any way except by the most calculated and cautious steps is politically fatal.

So leaders display a range of styles. Some have to do with the way business leaders deal with or manage others, some with the way they use personal style to get things done. In either case, it is possible to capture the continuum of styles in visual display (Figure 5-1). This is not the most detailed look, but it does demonstrate the range of behavioral choices available to leaders and managers.

From tyrannical to relinquishing is a long way. The tyrannical manager is impossible to manage or deal with; he is completely self-absorbed and out of touch with the emotional needs of his constituency. At the same time, whatever motivates the relinquishing manager to ignore her people is equally difficult to live with. The positions between authoritarian, your average hard-to-deal with boss, and delegating, the one who can't make a decision, are more common.

Over the years we have been training managers and leaders to understand the effects their approaches have on worker productivity. We have looked for helpful tools, models, and indicators, and have found many that can help managers and leaders perceive their own styles. One, the Self-Awareness Indicator, is used to show managers and leaders how their own perception of themselves contributes to their success with others. The indicator itself is rather lengthy and requires a facilitator to complete, but the

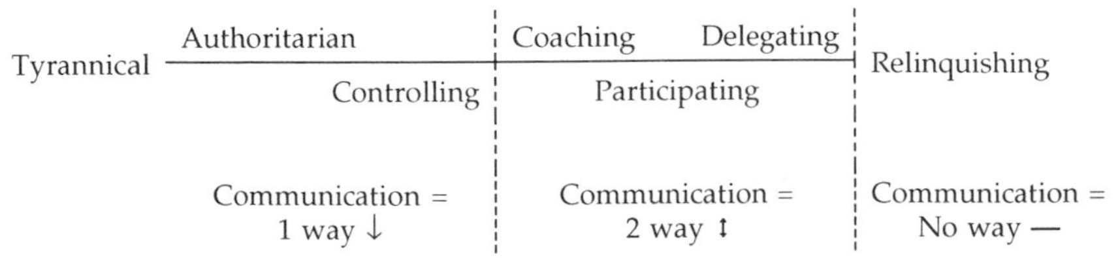

Tyrannical
Authoritarian
Controlling
Coaching
Delegating
Participating
Relinquishing
Communication = 1 way ↓
Communication = 2 way ↕
Communication = No way —

FIGURE 5-1. The Continuum of Management Styles

Brain dominance models lay out various modes of thinking preferred by left- or right-brain people. These models, loosely based on physiological models of the brain itself, contend that our work and thinking styles are related; that is, we tend to gravitate toward work that reflects our preferred style of thinking. Therefore, a person who prefers left-brain thinking might become an accountant or a computer hardware engineer. A right-brain-dominant thinker, however, might be happier as an organizational consultant or a visionary entrepreneur.

The style of communication and management displayed by people with various brain dominances might move predictably along a continuum of thinking styles. In broad and general terms, the more right-brain dominance, the more sensitivity to people we would expect the leader to demonstrate. The more left-brain dominance, the more technical the person's orientation and the less likely that person might be to demonstrate a more personal, empathic capability. In other words, the left-brainers have to work harder at empathy, and the right brainers need more help with analyzing the numbers!

More classic psychological approaches investigate ingrained and learned behaviors and how people have translated them into repeated patterns of behavior. Some psychologists would say these are emotionally driven, others that they are conditioned responses, learned by trial and error and success and failure over time. The emotional origins of styles are based on the complex patterns of personality development and the lessons learned throughout our lives to satisfy our inner needs and expectations about the outside world. The origins of these more emotionally driven motivators are hard to retrace and understand, so deeply encoded are they within our neurology. The behavioral approach has been more attractive because it is simpler to understand: changing an outward behavior to become a more effective manager is a feasible undertaking in a work environment; retracing one's life to childhood to uncover the roots of a Type A work style is not.

These more classic psychological models are more complex, but they remind us that habits are hard to break. Managers or leaders who really expect success at learning situational flexibility and increasing their empathy with others will have to work at it with some diligence, unless it is already a part of their personality.

We encourage managers and leaders in the technical world to avail themselves of training and development opportunities in this area, since identification of one's own style helps understand the affect we have on others. In addition, the better examples of such developmental activities use practice techniques to assist individuals in applying new skills that can later be translated to the real management world. Each person must find the approach to skills enhancement that works best for them. We encourage pursuing training—whether based on brain dominance, behavior, emotional predisposition, whatever it takes to help one understand oneself and learn to inhibit or support those aspects of oneself that enhance leadership and management effectiveness.

ASSESSING DIFFERENCES BETWEEN MANAGEMENT AND LEADERSHIP

Management is generally thought of as the ability to get work done through others. It means managing tasks and projects, harnessing resources, keeping track of budgets and activities. Managers define work and organize the environment to make it easier for others to do it. Managers are called on to give feedback to others and provide encouragement and support. They are positioned to pass communication up and down the hierarchy and to be sure people know what they need to know all along that structure. Managers report on progress, track quality and output, and direct the operation of whole departments and divisions.

Leaders, on the other hand, inspire. They rally troops to causes and show the way. Leaders emulate the best way to do things. They reach into the unknown future and articulate vision. They hold to the big picture, leaving the day-by-day to the managers. Leaders are supposed to be bigger than life, seeing further into the future they dream for us. They imagine what others can only guess at, and see clearly the path to follow.

Leaders are supposed to display the characteristics of success. They embody some goal or achievement to which the rest of us aspire. Reverend King was a man of peace, John F. Kennedy a political hero; each overcame great odds to further his cause.

The characteristics of leadership can help the manager to excel. She must excite people about the work to be done, getting them to commit to the tasks essential to complete the project. More and more the manager is called on to articulate the vision of the department, even the whole organization. The manager as leader rallies the forces to a cause that perhaps only she truly understands until she finds a way to convey it for others to digest and appreciate. The manager must model, like a leader does, to lead others to change, grow, and excel.

Leaders may not have to do much management, at least that we see, but more and more managers are called on to lead. This trend is likely to continue. So will the trend to debate the differences and similarities between management and leadership. Whichever way the debate unfolds, leaders are essential to our future and to our ability to see our way there, but without management the best vision is useless. Both are vital to our success.

THE MEANING AND CHALLENGE OF LEADERSHIP FOR THE MODERN AGE

Things are changing drastically in the 1990s, and the pace of change can only accelerate. As the world reels with emerging political alignments, evolving economies, and an ever growing and demanding population, we grow nearer and nearer to some new brink. The lesson of the quantum scientists to the organizational leaders is that every system has its breaking point. Chaos is a millisecond away from order, and crossing the boundary is increasingly easy as systems become more stressed.

Our businesses and organizations seem stretched to their limits. The pressure of advancing technologies and the information overload they bring, along with the stresses and strains felt by a collection of average people trying to make their organizations work, adds to the strain. Huge organizations and the stability they offered for so many years are now imploding. Completely new ways of working and new kinds of services and products are surfacing in a rapidly changing and increasingly complex infrastructure. What the individual did yesterday to succeed at work and bring home a paycheck may cause his failure tomorrow—whether he's a CEO, local manager, or department head. No one is left untouched by the post-industrial, post-information age convolution we are experiencing today.

From everywhere we hear the advice to learn quickly, stay agile, and embrace the idea and practice of change. These skills will characterize the winners, and the losers will be those who cannot adapt. Creativity is the vital skill of the changing millennium, and leadership is in demand now more than ever.

Both the individual and the organization are called upon to change. Leaders are needed to show us how to combine individual success and excellence with expanded teamwork. Somehow both must evolve at once.

Creativity and leadership. Two powerful and elusive qualities on which we are pinning the hopes of the business world. As organizations and working people, we have put out a call for the leader of tomorrow, and we are accepting résumés. We cry for leaders who can think differently and expand our perceptions of how things are and could be. These leaders can see far into a future that remains bright and alluring in the face of great ambiguity and uncertainty. Every organization seeks leaders who can show us how to distribute power throughout the company without losing focus and direction. Leaders are needed to model integrity and courage in the face of vast changes, and to help us make the right choices as we build the future. And leaders are needed who can infuse organizations with mission and vision while making a profit in the hard competitive environment of the 1990s and beyond.

Finally, the leader we seek will help us deal with our cultural differences and our need to be unique, while enhancing the environment for successful teamwork. All of this while remaining the model of calm, staying spiritually centered, and making a profit.

No one person can do all of these things. We need a climate in which we empower and foster leadership in everyone. Managers, line workers, CEOs, all need to rethink what we can do in our own ways to become more effective leaders.

At the beginning of Charles Dickens' novel, *David Copperfield*, the title character says: "Whether I shall turn out to be the hero of my own life or whether that station will be held by anybody else, must yet be shown." The best hope we have is that our future leaders will help everyone become heros of their own lives by modeling a new work ethic where both teams and individuals work to build a more creative and empowered environment, ultimately allowing all of us to be enabled leaders of our own destiny.

6

CHAPTER

MANAGING UPWARD: HOW TO DEAL WITH YOUR BOSS

THE REALITIES OF THE REPORTING STRUCTURE

Hierarchies are in place to get work done. Ours is a system of distributed authority that has evolved over time to streamline decision making while balancing centralized and decentralized management. Although some aspects of that structure and hierarchy are under attack today, the chances are pretty good that you report to, or are accountable to, someone else besides yourself in the organization where you work.

What does it mean to be a boss these days? Bosses are people who play a role in the structure. They are given positional power so they can carry out the assigned work of the organization through the top-down authority of the hierarchy. As we will discuss in Chapter 9, "Managing Without Authority," this does not always assure success since power is derived from many different sources. Nevertheless, the positional power enjoyed by bosses does lead, in the worst of cases, to a good deal of conflict, envy,

A CASE STUDY

Tom and Carla had worked for several years under the same R&D manager, Martha, at their research laboratory, which was part of a larger technical organization. They had often talked about the stress and strain they felt working for her because she was difficult to communicate with, managed through intimidation, and generally treated her staff as if they were never quite good enough. Both Tom and Carla felt that Martha withheld vital information, and that she had no idea how to motivate and support the staff, let alone keep them up to date on technical and organizational events. Tom grew very depressed during his time under Martha's direction. He had become suspicious and had lost all but the most cynical aspects of his sense of humor. He came to work late now, and left early, often leaving unfinished the documentation of projects he used to care about.

Recently, Martha had moved, or been moved, out of the position. Rumor had it that her upper management had grown exasperated with her and had moved her to another area of the organization for closer monitoring. Currently, Tom and Carla found themselves working for another boss, Harry, who has also challenged their patience at times.

Carla, remembering what life was like with Martha, has been trying to learn from it. She has been going out of her way to listen to and ask questions of Harry, to find out how he thinks and communicates. She has learned that he likes his information in carefully written summary memos that document in words the state of the business. He makes decisions based on information passed to him from his team, and expects that information to be timely, accurate, and easily accessible. Harry is annoyed by long, rambling meetings and by people who bother him with too much detail. That isn't to say he doesn't value the facts: he does more that anyone Carla has ever worked for, but he wants them succinctly presented in writing, and he wants to see them followed up by at least two and preferably three interpretations, along with the same number of recommendations. He also expects a description, in the form of a bulleted list, of the potential implications of the recommendations offered. This helps to speed up his own thinking and facilitates his decision making.

Finally, Harry's management style is upbeat, and he doesn't hear bad news well. He is optimistic himself, and his presence is always good-natured and positive. This is actually infectious to those around him, who can get through his somewhat formal veneer to see how he really likes to work.

Carla found out all this by asking him, observing him, and by some trial and error. She was surprised how straightforward Harry had been about what he needed for his own communication style. Unlike some people who might panic when someone in charge of them asks for information in a format or style they are not familiar with, Carla set out to adapt her own more anecdotal and open-ended style to her new boss's style. She could do it partly because she knew it was in her best interest, partly because she is a flexible person who prides herself in her own agility, and partly because she didn't take it personally that Harry was different from her and that he happened to be her boss. She had also learned a lot by suffering through her experience with Martha. She often said that she would be a better manager herself in the future because of the negative role model Martha had been. Carla, unlike Tom, had kept her positive energy throughout and was not deterred or embittered by the experience. On the contrary, she was looking forward to moving on to better times ahead.

After a while, Carla had an opportunity to test her skills at working with her boss. She had received news from one of her technical research staff about the availability of funding to investigate a controversial area related to her group's research. The money would be hard to get, requiring some front-end proposal research and writing time. In addition, the technology under investigation was a bit controversial and not exactly in the mainstream of most of Carla's work, but she felt compelled to pursue it. She felt that even though the technology had only a 5 to 10 percent chance of leading to an application breakthrough, it would revolutionize their field by 10 percent. The odds were long, but ignoring the opportunity seemed riskier to her than spending the time and political currency to try for her management's support.

For several weeks, Carla worked quietly behind the scenes with her staff to put together a proposal and to document carefully the pros and cons of the opportunity. They researched the sources of available funding, investigated potential competitors, studied and documented the costs, risks, and potential outcomes of the research at various levels of success and failure.

Then Carla went to speak with Harry. She scheduled half an hour with him, and arrived with a five-page executive summary of her ideas, including three clearly produced charts to capture her argument in numbers. She presented her case succinctly and with enthusiasm. She looked excited about the possibilities, and although she acknowledged the risks, she did not dwell on them. In short, she presented a balanced, but upbeat view of the challenge. Harry was impressed by her clear presentation and carefully worded and produced document. He promised to study it carefully and get back to her by the end of the week.

Harry did that. On Friday morning, he called Carla and asked if she could put together some overheads based on her document, including the charts, to present to him, his two superiors, and a member of the accounting department on Tuesday of the following week.

On Tuesday Carla gave another succinct, easy-to-understand summary presentation of her original document with easy-to-follow overheads, and answered several difficult questions. Harry and the three other managers were clearly impressed and promised a decision within two weeks.

Harry informed Carla that they were willing to get involved if she would continue to keep them informed in the manner of her original presentation. Carla and her technical staff were ecstatic. They were launched on the most exciting work of her tenure at the organization.

Tom, on the other hand, was so disturbed by the relationship with his previous boss that he could see nothing but another bad experience in the demands he felt coming from Harry. Tom recognized the closed "do it my way" approach to work that had characterized Martha, and he assumed he was in for more of the same. After all, hadn't Harry announced quite dogmatically on his first day just how frequently he wanted reports and in what format? Where is the room for creativity, personal expression, or real communication? So Tom was defensive from the start. Tom was not a person who enjoyed the concept of a boss anyhow, and to have another potential tyrant in the job—why, it made Tom tired just thinking of it.

Tom's first act of rebellion came at the first reporting date. He laid his report out in a format altogether different from the template provided by Harry and attached an explanation. Tom indicated that the format provided by Harry didn't allow Tom to provide the big picture, and that too many crucial details of his work had to be omitted to adhere to the guidelines offered. Tom made a point in his cover memo about not wanting his personal creativity stifled by the bureaucratic procedures Harry was setting up.

Under Martha, Tom had given up trying to persuade anybody above him of anything. But the situation in his department concerning the availability of computer equipment was becoming acute. Tom's inability to convince Martha that his group needed more technical support had put them in a precarious position for resources. Pressure was mounting from Tom's direct reports to get them some help. One of the members of Tom's group had brought his own personal computer from home as a last resort.

Tom girded up his energy for the fight ahead. He pulled out extensive time reports from on-going projects of all kinds and spent hours writing up a detailed description of how far behind everyone was and how miserable working conditions had become. He wrote and wrote, documented and documented. After several weeks of building his case, Tom had over 100 pages of solid documentation about the difficulty of meeting challenges, deadlines, and quality standards because of the missing equipment. He showed in great detail how each member of his staff had been frustrated and set back by earlier upper management neglect, and how more of the same would lead to disaster.

Finally, he finished this plea for help and sent it by interoffice mail to Harry. And then he waited.

And waited some more.

Finally, after several weeks and his staff asking about upper management's response, Tom told them that he had tried and tried, but that no one would pay any attention to them. He told his staff about his great report and how the narrow-minded bureaucrats in Harry's world were too out of touch to care and hadn't even gotten back to him.

Harry had seen Tom's 100-plus-page report, but couldn't make anything of it and hoped Tom would let him know somehow what he really wanted or needed to manage things better. Harry had passed the report to someone on his staff to review, and whoever had it had not done so yet.

Finally, word got to Harry that Tom was having trouble managing his group, and that people were upset by how Tom managed and secured resources. He called Tom to his office to find out what was going on. Tom went into a long harangue about how his people were at their wits' end being ignored like this, and how he just couldn't be successful without an infusion of resources immediately. He left the meeting with Harry in a huff, convinced he had made his point by "being strong."

The next day, Tom received a memo from Harry saying, essentially, that now was not the time to invest in further computer systems for his group. That instead, Harry was sending in members of the accounting and computer services department to determine the real issues and what was needed to help get his staff back up to acceptable productivity levels. Harry also indicated that Tom might want to visit the Human Resources Department and see what kind of management training they had in communication skills, Harry's memo said, "to improve your interaction skills with your staff and others."

A Short Analysis of the Case Study

What happened to Carla and Tom in this story suggests several questions. Why did Carla achieve so brilliantly? Why did Tom's efforts not only fail, but lead to an investigation of his whole operation?

Clearly, Carla's position was enhanced by her ability to back off from her own emotional responses to things and analyze what was needed. Instead of taking Harry's behavior personally, she merely noted it as a style to contend with. She recognized her own responses to Martha's tyrannical management style, but didn't let that be her driving motivation in responding to Harry. She saw it as a problem to be solved: "How do I get someone who thinks and works very differently to see the challenge as I do and understand what needs to be done?"

With that approach, Carla freed herself from her own and her boss's emotional triggers. She could then build a logical persuasion process using the proof material she knew Harry would respond to and packaging it in a way he could address and understand easily.

Proof material is the information and the way of organizing it that lets others understand in their own way the situation and its potential solutions. An analytic thinker sees proof material as a set of numbers or data that add up to a certain course of action. A more interpersonal approach might require a set of proofs that showed how a course of action would directly benefit the people involved. For that way of thinking, the well-being of people might be more important than the short-term effect on profit.

Another proof set might include the affect of a certain coarse of action or decision on the business after two to five years. These proof sets are very different for different people. Carla saw that Harry's proof set needed to offset opportunity and risk in a calculated way that included financial and technical data. So that's what she gave him.

In understanding what happened to Tom, we have to see the emotionalism he brought to the process. The proof set Tom presented was his own, not Harry's. Furthermore, Tom's proof set was about how badly he perceived he had been treated, not about how business and productivity would prosper by meeting his management request.

Carla did better with a risky new venture than Tom did with an obvious need for remedial technology because of the ways each one of them managed information, communication, and relationships upward. Carla was adaptable and stylistically flexible. She packaged her information in Harry's language, format, style. Tom allowed his emotions to dominate and produced a confrontative communication that led to an investigation of his entire operation.

■ ■ ■ ■ ■

In the next section, we look at the rules and behaviors for enhancing upward management. Examples of each of the following points can be found in the case study.

MANAGING SUCCESSFUL RELATIONSHIPS WITH YOUR BOSS AND THOSE ABOVE YOU

☞ *Accept that your boss's support is important to you.* You really can't do it alone, and you would be arrogant to think you could.

Anyone who sets out to compete with the boss had better do it with a sense of danger. Not all bosses are good at getting you the support you need, but your challenge is to help them find ways. Even the not-so-good ones can get more done with your help than you can without theirs.

☞ *Understand how important your support is to your boss.* The boss needs you. Her future depends on your success on the job. You know things and have access to information she needs to be successful. Your openness and willingness to share that is your key power base with her, and managing that power base openly, but carefully is your trump card. Additionally, you undoubtedly have skills your boss desperately needs in her job; presumably you wouldn't be there unless you could do certain things that are crucial to the organization's work. *Keeping* your skills up to date and crucial to your boss's success is your best competitive strategy.

☞ *Understand your own response to the boss's style and personality, and manage it.* We all bring certain psychological baggage to the work place. Some of us are still mad at our mothers and fathers for the way they used authority over us. Some have never recovered from the largely disastrous teaching styles forced on us in school. These and other experiences can drastically shape our responses to those with power over us today. It's important to know your hot buttons so that you don't act out old scripts in your current job. Get to know your own psychological framework and triggers, and be aware of them. Then do what you can to keep them from sabotaging you like Tom's did in the case study.

☞ *Respect the style and orientation of your boss to his work.* Let the boss be who he is; don't try to change him. Spending your whole life complaining about him diverts your energy from your work to a hopeless cause. Focus on your own success by supporting what you know needs to be done and what makes the boss look good. Forget changing the style of others—you can't. Learn to value those other styles as expressions of human uniqueness. We know of a human resource manager who spent so much time trying to change her boss, in this case the president of the company, that she could never even stop and acknowledge the progress and success she had made in her own work. So the boss never gave her praise. The rest of the company did; she should have celebrated that and moved on. Instead she wallowed in the depression of being unappreciated by her father, excuse us, her boss. Don't confuse the style of others with your own baggage—or your own goals.

☞ *Understand your response to your position in the hierarchy and how you feel about working within a structure.* This is another area where our psychological baggage can bog us down. Be aware of it, see a therapist if you have to, but don't let it block your willingness to succeed. Getting ahead for getting ahead's sake may be your aspiration, but challenge it. Remember, power is a commodity for getting work done. As a goal in itself, power is always disappointing. Any seasoned boss will tell you: power as a route to achievement is fulfilling; power as an expression of mindless control over others is finally debilitating.

☞ *Learn to take feedback objectively, not personally, and maintain your sense of self and your own uniqueness.* Everyone needs and everyone dreads feedback, sometimes even when it is positive. Being told how good we are can be as embarrassing, and carry as much pressure, as being told how much we could improve. But the trick with your boss is to *ask* for it, work with her to learn from it, and ask for guidance in how to improve based on it. Then, ask for more. Bosses are not so good at feedback. So help her not to surprise you at review time by being open to feedback all year long. Who knows, the boss may ask you for some, too. Good ones do.

☞ *Push back when necessary, but for business reasons and to maintain personal integrity, not for political gain or to embarrass the boss.* While you ask for and give feedback, remember that position does have its perks. Go for what you believe is right, but do it for reasons that support business goals and out of vigilance for your own integrity. Making the boss look wrong or like a fool may give you limelight for the moment, but it is seldom a good career-building move.

☞ *Learn the boss's goals, aspirations, frustrations, and weaknesses.* This you do to support the boss, not to undermine him. By learning what his goals are, you learn how to manage your time. Your A priorities are his A priorities. By learning the boss's dreams and aspirations for the future, you help him reach them, and in so doing make opportunities for yourself. By learning his weaknesses, you can cover for them and make him look stronger. All of these behaviors are to your benefit.

☞ *Study and understand what the boss thinks is important: what is her proof set?* This was a key issue in the case study. What the boss thinks is important can be determined by what she uses to measure success and results. Use those measures only in trying to persuade her of something, especially if it's a new or innovative idea. Be aware that these measures may be less conventional ones like employee satisfaction, or more mainstream ones like contribution to the bottom line. Find out and use her measures to support your work and ideas.

☞ *Study and be able to emulate, for the sake of being heard, the boss's communication style.* Once while attending a cocktail party, I was cornered by someone I did not know who talked on about himself for half an hour. I just listened and smiled, saying nothing. Later the same person told my wife what an interesting person I was.

Listen to your boss. Find out how she thinks, what she thinks about, and how she expresses it. Emulate that style, focus on those issues, and enjoy the fruits of your empathy.

☞ *Be dependable. Follow through on serious requests for information and work output.* Nothing impresses anyone more than reliability. Your boss needs you to be reliable and rewards you for it. Sometimes you will have to use judicious inquiry and your best after-the-fact judgement to determine what your boss really wants done and what is just talk. When you have determined what counts, do it with quality, timeliness, and enthusiasm. Give a little more than the boss asked for, and don't wait around for praise.

☞ *Display to others and expect from others mutual respect in all matters of business and on-the-job interpersonal interactions. This means respecting time, resources, and alternative work styles.* Mutual respect is the hallmark of all successful on-the-job relationships. In the case of the boss, practice it scrupulously. Some bosses are not good at respecting their subordinates' time and need for resources. You are justified in taking such a boss aside for a private discussion. Mutual respect works both ways, no matter where you sit on the hierarchy. You should expect it to be offered you, no matter what race, sex, or life style preferences you pursue. It is just another version of the golden rule.

☞ *Be honest and share all relevant data about situations and concerns at hand.* It hardly ever pays to keep business information hidden from the boss. Sometimes the information is too small to bother him with, but even then it is generally better to err on the side of too much information (bearing in mind the context of what you know about his style of managing). Information withheld for power purposes will only backfire on you. Always share what the boss will be better for knowing. And the most important rule of all: Don't hide bad news from him. Find ways to get resistant bosses to hear it, and share it as fast as you share good news. In the context of the other rules here, you and your boss are problem solvers together, so reinforce that view by raising problems before they get out of hand and taking whatever part you can in solving them.

☞ *Keep private any criticism and conflict that may arise between the two of you, and always work for a jointly satisfactory solution.* A creative relationship with your boss is far more satisfying than an adversarial relationship, especially over time. So work on turning different points of view into opportunities to create new approaches and solutions. And never humiliate the boss in public, however upset or disturbed you may be. This is a major sin of organizational relationships, and you cannot recover from it. Unless the boss specifically asks for public discussion of how she might improve, keep your criticisms and suggestions between the two of you in a private setting.

☞ *Be manageable by and available to those beneath you. This assures you know what is going on and that you have the best information to pass upward.* To be of ultimate value to your boss, you must be a good boss yourself. Avoid the trap of managing as you are managed, unless your manager is a brilliant role model, in which case practice whatever you can under his tutelage. By being a successful boss yourself, you collect valuable information about the organization and what it is doing, and you oversee a successful operation of your own. Your boss needs good bosses to rely on, so put it on your personal calendar to avoid complacency and continue developing your management and leadership skills.

SUMMARY

The hierarchy exists to get work done, and the opportunities to succeed within that framework are wider than we sometimes think. Managing ourselves sometimes means managing the boss—an unlikely job description, but part of the work environment, nevertheless.

The key is to be flexible and objective. The lesson is to stay open to differences and manage for the best outcomes possible within every relationship. Beyond that we can only hope that the people we are asked to work for are reasonable and competent, and that we can learn from them as we build our careers and expand our networks of professional contacts.

7 CHAPTER MANAGING CONFLICT

In all organizations where people work together, there will be conflict. In and of itself, conflict between people is neither bad nor good. The effects of conflict are much more a result of how we manage it than of the conflict itself. To manage it more effectively, it is important to understand its sources and the choices that exist for handling it. By identifying and understanding these sources, we can better select the means of dealing with it most appropriate to the situation.

Conflict has both liabilities and assets for an organization. This comes as a surprise to some managers, who believe that the best way to deal with conflict is to wait until it goes away—a precarious approach, at best.

The liabilities of conflict include diminished productivity, job turnover, burn-out from dissatisfied and unhappy work experiences, project interference or collapse, gridlock and lack of decision making, communication breakdown, and generally a poor quality of work performance and output.

The assets of conflict include a potential for more creative output from groups and teams, a variety of experiences and interactions in the workplace, the opportunity to learn from other ways of thinking and working, new stimuli that motivate and hold interest in work, and last and most importantly, a large pool of differentiated resources from which to draw talent for various projects and work requirements.

SOURCES OF CONFLICT

The potential for conflict is everywhere. Its specific sources vary from time to time, place to place, and condition to condition. Some of the more commonly recognized sources are discussed here.

Organizational or Structural Conflict

The way an organization or a project is set up can cause conflict. In general, the very hierarchical nature of organizations causes conflict by encouraging various forms of competition. In the case of competitive situations like those we see in most work settings, competition has been implicitly sanctioned as a way to interact with others; the conflict is said to be built into the system. The expectations of individuals and groups in competitive conditions, like sales forces after the same business or research departments after the same funding, further defines the origin of intergroup and interpersonal competition and conflict. This is not to say that all such competitive and conflicting arrangements are bad, but rather to define and isolate a structural source we need to be aware of in assessing conflict's causes.

Another source of conflict at the organizational level occurs between departments and groups established to carry out different functions. The people who act as liaisons between such groups see the most evidence of the conflict. For example, the engineering department has a liaison to the manufacturing department. That liaison sees the bulk of the conflict between engineering and manufacturing as technical specifications are turned into manufacturing procedures. The practical efforts of the manufacturing department come face to face with the idealism of the engineers, and we know what happens next.

Another example of the conflict that occurs at such junctions is the relationship between the purchasing function and nearly everyone else. The inventive engineer who needs resources sees purchasing as a nightmare of bureaucratic paperwork that stifles creativity and enthusiasm. Purchasing, in turn, sees the inventive engineer as a spendthrift who can't plan ahead. This conflict very often escalates into harsh words and heavy artillery.

These are only two examples of organization-based conflict between groups. This occurs primarily because people do not readily see the work from others' points of view. When driven to perform in our work environment at the tasks for which we are responsible, our entire effort is focused on the goal. We strategize constantly how to marshal the resources to succeed, and everybody else can look like an obstacle. From this perspective, it is easy to fall into conflicting and adversarial interactions. The obvious nagging realization is, of course, that we would be more successful at garnering support from others to get our work done if we could find ways to bridge the structural gaps between us and make contact on some other level.

In an effort to do just that, I once sponsored a "Take An Engineer To Lunch" program at the computer research and manufacturing company where I worked. The program reimbursed two or three people for their lunch bills (once a year) if they submitted the receipt and a five-line report telling what happened when they invited strangers from another department out to lunch just to get to know them. Follow-up indicated that productivity and relationships between the departments improved as people became aware of each other by name and personality. Instead of "fighting with purchasing," it became Frank calling Alice for help. This revolutionary idea of taking a stranger from another department to lunch points out the value of overcoming the anonymity between people of different functions separated by organizational structure. It's harder to lump someone into a group of undesirables if you know him by name and have had lunch with him.

This conflict between members of different departments and functions extends to whole organizations, as well. One of the biggest problems faced by organizations that merge with other companies is the conflict of the two cultures. It can take years to work out those conflicts. In the case of a large and well-known,

U.S.-based pharmaceutical company, the conflict between merging subsidiaries nearly destroyed the ability of the whole company to compete. To this day, conflict of purpose and of policy, both perceived and real, inhibits the company's growth and competitive market position.

Role Conflict

Role conflict is at the core of what happens between people in competing functions and departments. Role conflict can also emerge within work groups and teams. This kind of conflict arises at the boundaries between jobs. Where my job ends and yours begins is an irritant for a number of reasons. First, we tend to define ourselves and our value to the organization by our expertise and skills. It's possible to be threatened when others display the same or even higher levels of skill at the work we do. Our identity as a person and as a member of the organization is tied to our job definition and requirements, so it is understandable that our first response may be to feel threatened when our identity is threatened, real or imagined.

Second, we tend to feel territorial about our areas of responsibility, thinking of it as our own empire of success. After all, we are paid to do this work! When others challenge that (or appear to), our personal security and sense of well being can be threatened. This can lead people to draw their guns to protect their turf. Protecting turf tends to blind us to the overall goals of the organization and shrinks our world view, but it is an understandable response. Frankly, in the changing and ambiguous 1990s, these feelings of exposure and uncertainty about the sanctity of one's turf are bound to grow, not diminish.

Other role issues may surface as well. This is especially true as organizations shrink, move toward team-based working structures, and integrate many diverse nationalities and ethnic backgrounds. The roles we used to play in the company may become less important, outmoded, or even replaced by newer models. Unless we as individuals have the personal agility to change with these pressures, and therefore change our roles voluntarily, we do face the possibility of being left behind. All of this can produce certain defensive behaviors that can show up in our attitude and performance on the job. Others experience us as causing conflict as we protect what we perceive to be our place, our role, in the organization.

Related to this are the requirements of the role itself. In any job you are paid to accomplish certain tasks and achieve larger goals. At times, the goals you pursue may appear to be different from others in the organization. Since you may all be competing for the same resources (people, time, money, even management attention), it's not surprising that conflict can result. The job, in relation to other jobs, is a conflict starter. Of course, part of the manager's task in managing projects and people is to find ways to share and optimize resources to minimize this as a source of conflict.

Another conflict that arises around roles is the conflict of authority, or in some cases the lack of it. On many project teams, for example, the lines of authority both within the team and outside it may be vague or badly defined. The project manager may have little or no authority, and those in the project may feel unsupported or as if their group has no clout in the system. A project manager with little hierarchical authority, who can't acquire resources and support through her influence, may unsettle the project team members and thereby induce a series of conflicts between herself and her teammates, as well as between the team and the rest of the company. This also leads to competition and disagreement over decisions and decision making, which produces another version of intrateam conflict.

Personality and Interpersonal Conflict

Sometimes the chemistry between people just isn't right. For reasons we cannot explain, we just don't hit it off. Any of the other causes of conflict listed here could be at the root of this, or the origins may be less tangible—either way, the conflict is just as real. Perhaps one of us is more left brained than the other; that is, more systematic in our management of time, money, or logistics. Or perhaps someone is more right brained than the main culture, and that annoys others; this may show up as wanting to work odd hours instead of the standard 9-to-5 pattern.

Matters of taste and style can also affect our interpersonal relationships and cause conflict. Even as trivial a thing as how someone dresses or answers his phone can cause someone else to

ceptions about what is right to do and wrong to do, what makes sense and what doesn't, and even who is smart and who is not.

The acceptability of an idea, and of the way of thinking that spawns it, is tied to our view of what is conceptually sound and what ideas are "good" or "bad." Therefore, we play devil's advocate for the ideas that don't fit our conceptual framework, and call people "smart" whose ideas are much like our own.

Conceptual conflict has a very positive role to play in our efforts to be creative and innovative, as will be discussed in Chapter 10, "Creativity and Innovation." In fact, of all the forms of conflict, if well managed, conceptual conflict is best encouraged in scientific and technical settings. It allows us to see relationships we would not ordinarily see and thereby to form new insights we would not ordinarily form. From these juxtapositions can come great breakthroughs. Ideational, conceptual conflict, understood and managed, is our single greatest human asset in group work settings.

MANAGING CONFLICT

Figure 7-1 shows an approach to managing conflict that encompasses a wide range of possibilities. Note that the columns represent various degrees of conflict within a working group (a team, department, or even a whole company). The rows represent three different management approaches to conflict along a continuum from avoidance management ("Suppress") to reactive management ("Resolve") to proactive management ("Celebrate"). As we move from the bottom row to the top row, we see the management of conflict becoming more proactive and positive. As we move from the left column to the right column, we see the degree of conflict increasing from negligible to intense. The figure shows the resulting relationships between degree of conflict and quality of management. Each of the cells describes these results.

The two differentiating factors between avoidance management and celebration management seem to be a willingness to engage in conflict and an expertise at dealing with the complexities of conflict. In short, when the management approach is characterized by an unwillingness to engage in conflict and an incapability of dealing with its complexity, the conflict is suppressed. Concur-

Management's Approach to Conflict	Low	Moderate	High
Celebrate conflict	■ Few problems ■ Environment open to conflicting values and beliefs	■ Increased energy and performance ■ Conflict seen as energizing	■ Maximum output of creative ideas ■ Conflict welcomed and encouraged as part of the creative process
Resolve conflict	■ Few problems ■ Environment equipped to handle diverging views	■ Increased focus on maintaining order ■ Conflict seen as tolerable	■ Crisis management to maintain order ■ Conflict seen as major problem
Suppress conflict	■ Few problems ■ Environment threatened by differences	■ Increased stagnation ■ Decreased performance ■ Conflict not acknowledged	■ Backsliding ■ Noticeable performance problems ■ Conflict denied

Degree of conflict

FIGURE 7-1. Approaches to Managing Conflict

rently, a slip in performance and obvious backsliding of the group's output and quality are noticeable. When the management approach is characterized by a willingness to engage in conflict and a capability of dealing with its complexity, the move is toward celebration and creative output.

Building on this, the manager is called on to broaden his scope and skills in dealing with conflict in the workplace. By opening up to the positive role conflict plays in work settings, he can extend his willingness to engage it. By enhancing his skills at dealing with the complexities of conflict, he can become more effective at getting positive results from the conflict management process.

One other aspect of conflict management is implied by the model in Figure 7-1. The model shows how managers tend to respond to conflict according to their own preferences and comfort levels. Some may constantly engage, others constantly avoid. However, at times, any of the management approaches might be situationally appropriate. For example, it is possible that suppressing conflict might make sense under certain circumstances, even if the manager is highly willing and capable of engaging at the celebration level. At other times, the manager who prefers to avoid conflict may see a value in collaborating with another party to achieve maximum satisfaction for everyone concerned. Rising above one's own preferred mode and demonstrating the flexibility to respond to the moment is the real lesson of the model in Figure 7-1. These techniques are discussed in more detail below.

In summary, by adopting more proactive attitudes and applying some specific conflict management techniques, the manager can greatly enhance her success at dealing with this daily challenge. Moreover, by understanding the range of options open to her in facing conflict situations, the manager can choose those most appropriate to the situation.

Suppressing Conflict

The manager who sees a situation in which conflict should be avoided or suppressed may accomplish that in several ways.

Direct Suppression of the Conflicting Parties. This means actively inhibiting your people from taking part in conflict behavior. This may be appropriate when managing a crisis or when there is no time or no cultural acceptance of conflict. In this case the manager keeps those in conflict on the sidelines or off the project until the crisis is past or the cultural conditions change. An example of a cultural condition is a disagreement between software engineers in the presence of a customer. This is usually an inappropriate place for them to demonstrate their clashing opinions. Another appropriate time to suppress conflict may be in a dire emergency. The captain of a large passenger jet about to experience a mid-air collision is unlikely to solicit alternative opinions about what should be done. Immediate action is called for on the part of the main decision maker. These are appropriate conflict avoidance situations.

Accommodation to the Other Party. Giving in to the needs of the other party may be your best course when the conflict isn't worth engaging in, or when you want some leverage with the other party in the future. The manager may also accommodate because he suspects he's in the wrong, or that he'll be unable to overcome the other party and has no chance of winning. He may lose more in resources by fighting than he could achieve by settling.

Resolving Conflict

Competition. This is when it makes sense to go for the win. Usually a manager will choose this option when she knows she is in the right. She may also choose this option when she believes those working with her will not be able to work without what she's fighting for, and there is no way to gain cooperation from others. This is a last-ditch effort to gain a needed facility, allocation of funds, or some other vital resource.

She may choose this option when to do less would put herself and her efforts in jeopardy from those who take advantage of a more cooperative style. Why try to be cooperative with those who will use your cooperative attitude to get all they can, and give nothing?

Negotiation. Volumes have been written about this best known of conflict management responses. Negotiation is the best selection when compromise is the best route to resolution. This is an especially attractive option when each side needs something from the other and both sides have excess.

Another time for negotiation is when you are gridlocked: compromise is the only option because no side will give an inch and each party has the power to impale the other. We see this in the kind of bargaining that goes on between labor unions and management.

Sometimes we negotiate for short-term solutions to problems or conflicts that are extensive and complex. It may be too difficult to discuss all the details of a project's resource allocation needs, so we settle for a temporary, less-than-optimum way of managing them just to get on with the project, or for a quick infusion of money or materials.

There are as many approaches to negotiation as there are people doing it. There are hard approaches that advise you to lie, to hold out, and to maneuver for advantage at every turn. There are softer approaches that encourage you to go in without a definite unmovable objective and be open to ideas and input from the other party. The range is broad, but the style is still to compromise for best advantage. The technical manager would benefit from further reading and personal training on the options available for successful negotiation strategies. Understanding appropriate times and places to use the negotiation approach is the first step.

Problem Solving. This is a more interactive version of negotiation. In this context, "problem solving" refers to a straightforward approach in which each party gets a result they can live with. Usually this kind of problem solving is technical in nature and deals with short-term problems or conflicts. Getting you to do X so I can do Y, making trades, and fixing some process that doesn't work to get more cooperation are examples. One specific case of problem solving came from the purchasing department example used earlier. Engineers at a large R&D company were not following purchasing's requirements in acquiring materials for their work. So Purchasing was delaying orders. A conflict resolution session between members of both departments resulted in a redesigned "Request for Acquisition" form that engineers could more easily understand and fill out. This solved the bulk of the problem.

Celebration: Collaboration, Creative Problem Solving. The celebration mode is a completely different approach to managing conflict. It assumes that we can always benefit to some degree from the conflict that emerges between people, groups, departments, and whole companies, if we see the value inherent in all points of view. Its primary orientation, therefore, is creative. Collaboration means getting everything you want out of the conflict and committing yourself to helping the other person get whatever they want. This strange sounding algorithm calls on our most creative capacities.

Collaboration is beyond problem solving, it is *creative* problem solving. It forces us to use all the skills and mindsets present in a situation (as will be described in Chapter 10, "Creativity and Innovation"). Without such an orientation, we will slip back into negotiation and compromise. It means we are looking for the new, unexpected, highly original solution to the conflict; we will not settle for enabling the opposing forces to work around each other, but we want to make them into partners. It's ambitious and bold, and not for everyone or every situation. On the other hand, in an age where creative solutions are sought after and very valuable, it is probably an option too easily passed over.

Collaboration and creative problem solving are best used when a new and unique solution is vital to both parties. When two parties are facing loss of funding and only by combining resources and efforts can either succeed or survive, collaboration is in order. When each of two scientists hold one half of the solution in their test tubes and a solution is needed now, collaboration is in order. When you need the information and perspective of an individual or a group who holds a key to your project or invention, collaboration is in order. When all else fails, try collaboration! Before you try anything else, try collaboration!

CONCLUSION

The truth is that conflict is rampant in human organizations from the smallest family to the largest corporation. This is to be expected: individuals with a wide range of experiences, backgrounds, and thinking preferences are thrown together in high-pressure environments. Such conflict need not be destructive and unmanageable. By educating ourselves about the sources of our differences and the options for handling them, we can simultaneously move toward the more collaborative side of the model in Figure 7-1, and choose from the array of available responses to conflict it describes.

Some of the chaos experienced by the individual is excessively dramatic. Take, for example, this true story of a man working for a major computer company that is undergoing enormous changes. In one three-month period, he suffered two heart attacks and watched his wife loose her 27-year position at a major retail store. Soon after his job was phased out, and with it all his medical benefits. Since he had his two heart attacks before his wife found her new job, her new employer would not pick him up on a family policy, and he lost the medical benefits from his former employer after eighteen months. From a position of relative security, this man and his wife experienced unprecedented chaos in their personal lives and paid a high physical and fiscal price for the changes going on in the world around them.

Such rampant and destructive change affects entire organizations. The complacent staff of AT&T was shaken to its core when the company was forced into divestiture in the 1980s. People who depended on Ma Bell for life-long work and advancement, as well as after-retirement care and support, were shocked and angered by the sudden insecurity of their jobs and their lives.

In these conditions of unbridled change and challenge, people are called upon to respond with personal creativity and resilience. But the individual working alone is at a serious disadvantage. The changes experienced by organizations and their members today are the result of interdependent forces too great for one person working alone to buck, let alone control or even predict. For example, how does an individual respond when a large organization like Digital Equipment Corporation or Ford Motor Company decides to lay off thousands of employees? How can an individual make a difference when a vast multinational corporation suddenly loses its profitable edge and starts posting major losses?

In this context of chaotic social change and unexpected personal upheaval, individual ingenuity is seldom enough. Even managing personal life changes is pushing the boundaries of individual flexibility and resilience. It's not surprising, then, that organizations are responding more and more by forming teams of highly skilled and trained individuals to pool knowledge and resources to face the unique and enormous challenges posed by the changing world. Although teams are called on to share their expertise for more creative output in shorter time frames, it is perhaps ironic that more creativity is demanded of the individual as well, as he or she is called on to participate in these highly visible, extremely complex team challenges.

Specifically, individuals are called on to work in unprecedented ways in teams made up of extremely diverse colleagues, as efforts are made to assemble talent from all functional areas of the company to solve problems and carve out future business directions. In addition, these teams are formed rapidly and must respond quickly to gain a competitive edge and develop workable answers to the seemingly overwhelming issues facing them. This need for fast results in the context of complexity and diversity means the team must minimize barriers to its success and be able to capitalize immediately on exactly those traits and conditions, such as their own inherent diversity, that often keep teams from being successful at all. All this in a setting where these teams are expected to excel as never before.

The implications of teams coming together in these settings are also unprecedented. Old ways of working together won't do. Top-down hierarchies from military models flunk completely when highly skilled professionals come together for short stints to solve major problems. In one pharmaceutical firm, we watched while a broadly cross-functional strategic planning team, which had been chartered to aim the whole company into the future, literally self-destructed as old ways of working and thinking about working seemed systematically to destroy the project. More will be said about specifically how teams are called upon to challenge the entrenched views of organizational management when we look more closely at the team effectiveness continuum.

THE TEAM EFFECTIVENESS CONTINUUM

Groups work together in many different ways, and certainly not all work groups can be considered teams. At some point, however, an ordinary collection of employees working regularly together might begin to demonstrate "teamness." Determining the exact moment when a group working together becomes a team is a little like determining when a cylinder of steel with wings hurtling down a runway becomes an airplane, but the move from work group to team is an observable transition. Just as the airplane finally takes flight, so the team finally elevates itself beyond its definition of a work group. In the case of the airplane, all the ingredients are in place before the actual liftoff, and the flying is just proof that the necessary first steps were completed. The team follows suit, displaying success only after all the ingredients for "teamwork" are in place.

Teams—which are, after all, miniature organizations—are subject to the same pressures and processes as individuals or companies. They experience periods of chaos and creativity, just as does the social milieu in which they find themselves. Just as an organization without direction can fumble into chaos, so can a team.

So, in the midst of the chaotic moment, with new and difficult challenges erupting all around, teams are formed to respond. How successfully and with what results a team moves beyond the chaos surrounding it and the chaos it reflects within itself when it begins work depends on many factors. These factors combine to define how effectively a team operates.

The factors are the attitudes, values, and beliefs that are visible in the team's collective behaviors. It is possible to identify these behaviors and to isolate the characteristics and attributes of the team's overall performance along a definable continuum. The continuum acknowledges that teams face chaos and enjoy creativity as full-blown states of being, and that there are intermediate points along the way. Figure 8-1 shows how the continuum captures the various states of a team's effectiveness.

Many conditions add up to describe the state of the team at any given point in its life cycle, so the flow of this continuum is not linear. The arrows show movement from anywhere to anywhere along the continuum depending on internal and external conditions.

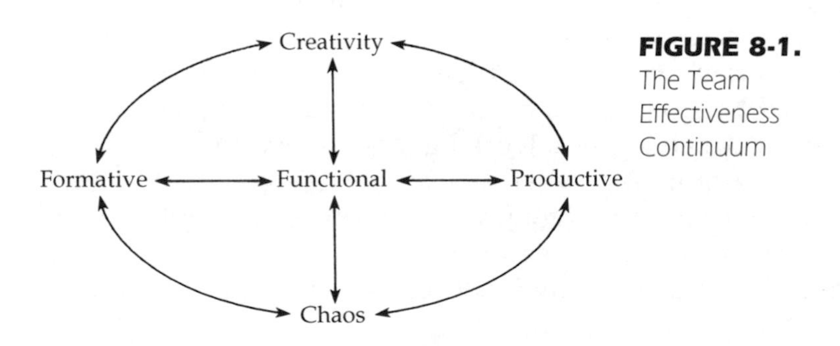

FIGURE 8-1. The Team Effectiveness Continuum

The Chaotic State

The chaotic state is exactly what the name implies. In this state there is no work group at all. Nothing from values to job descriptions is defined. No leadership exists, and individuals have no sense of what is expected of them.

Generally, chaos occurs when the organization is undefined, after a major reorganization, or after some calamitous event such as a major business failure. Teams may be sent into chaos by a parent company acquiring or divesting some part of the organization, or even when current team members leave the group for any reason. Adding a team member or merging with other teams can cause the same degree of chaos as other unexpected and disruptive events. Even positive change like the sudden funding of a highly desirable project can lead, if only briefly, to chaos.

Groups from any place along the continuum may experience chaos in the face of major and unexpected change. Chaos is an extreme stress response. Something has broken the accepted structure of things and caused order as we know it to crumble.

Of course, chaos offers potential advantages. A new and more appropriate order may emerge as the conditions inside or outside the team change suddenly. Out of the ashes of the previous condition can emerge a structure that will allow a new and more creative team to form. In this way, emergence from the chaotic state into the creative state, for however brief a time, is a very possible scenario. In extreme chaos, teams can pull together remarkably and invent a way out of the chaos that has all the hallmarks of the creativity response.

The Formative State

This state is experienced by work groups when originally forming or significantly redesigning their structure, personnel, or work definition. Groups changing the nature of the business they are in will also need to reform to face the new challenges they have taken on.

At the formative state the group is actively involved in defining or redefining its values, vision, mission, goals, objectives, and basic structure. Individuals will feel very insecure, without a sense of where they fit in or what the future looks like. Like the chaotic state, the formative state is scary and members may feel vulnerable and defensive. These feelings may manifest in searching for safe haven somewhere else, grabbing for power, or even dropping out and hiding in whatever way possible. During the formative period, information may become the ultimate power, and those who have it may use it as weapons in their ascent to leadership. In some cases, personal charisma outshines information power and leaders emerge from those who give most comfort to the stressed. Occasionally, new leadership arrives from outside the group altogether.

In the formative state, the group argues heatedly about the nature of its calling, mission, and who should be leader. The toll can be high for those unsuccessful at selling their points of view, and casualties are possible. In other cases, the formative process can be stimulating and exhilarating for all involved, especially if common interests and vision are promoted by interpersonal relationships and communication strategies that emerge and begin to work. Such a team may pass quickly from the formative to another position along the continuum.

The Functional State

The functional team has clearly established its mission and goals in concrete and specific terms, without much room for discussion or flexibility. Generally, wherever the vision, mission, and goals may have originated, they are now managed from senior management above the team. Individual team members buy into the preset and managed vision, mission, and work definitions as the last word, and think of this as perfectly acceptable and reasonable.

Functional teams get the work done. They are able to meet deadlines, specifications, and budgets. They follow guidelines and operational descriptions that have become well understood and repetitive procedures. The head of the functional team is a good manager, following the rules while ensuring that ongoing production requirements are met. The group has spirit and commitment to the tasks at hand. The enthusiasm displayed is based on the knowledge of how and why a good job is always done, and a belief in the constancy of the work environment, its expectations and rewards. The functional team, then, thrives on routine and predictability, and goes out of its way to reward consistency in output, attendance, and performance standards among its working members.

The Productive State

The productive team differs markedly from the functional team. This team begins to go beyond the expected norms of achievement. Motivated by a competitive drive to go one better than average, productive teams begin to find ways to share information to increase their output. They work hard to resolve conflicts that may set them back, and they expect leadership from the top in making decisions. Productive teams display pride in their teams and will defend them with a zealous patriotism. They often play intramural sports in competition with other teams in the company, and they demonstrate a moderate level of social bonding with each other.

The productive team maintains certain norms and standards that all must follow to maintain good standing. This team expects a certain allegiance to the goals, mission, and vision imposed from the top of the power structure. The question "How can we better achieve what is expected of us?" remains open. It is not unusual for the productive team to assign subsets of itself to work on special tasks to determine ways of improving the systems, procedures, and infrastructures that support the work of the team. This attention to the supporting systems is the beginning of looking differently at how work gets done. It is largely recognized in the current business context as an ongoing continuous improvement process, often called "total quality management." In reality, it is the first step toward teams taking charge of their own conditions, performance, and eventual output from beginning to end.

The Creative State

It is the creative team that has become a different breed of human work group. This team moves well beyond the initial awareness and interest in "quality," or continuous improvement, to investigate how to do it better. The creative team embodies all of the other states almost simultaneously in that it uses chaos for learning, forms and reforms as the conditions demand without losing track of its context, is highly functional when need be, follows standards of productivity consistently, and at the same time creates all new information, services, products, and meaning for others to emulate and celebrate.

This team not only shares all available information, *it is committed to generating new information as yet unsynthesized and to sharing it beyond the team. It is information porous.*

The creative team is probably self-selected, highly bonded socially, and committed to the success of each of its members. The creative team is committed to its vision because it has invented that vision. In fact, *creating its own vision—and therefore its own future and the future of those affected by it—is one of the creative team's key motivators.*

This common vision may be anything that brings its members together—a product or service to be invented and sold, a medical breakthrough to be achieved, or a peace treaty to be won. Whatever the vision is, the driving force of the whole team is its commitment to that vision and to the individuals assembled to achieve it.

The creative team rarely accepts a top-down assignment. Its work grows out of its self-selected vision. The team sees itself as the generator of its ideas, style, approaches, solutions, information, leadership, and values. The leadership that arises is shared and is more than a function of any given situation. Leadership in the formal sense truly withers in the creative team, and as a result of the incredible common commitment driving the team, a kind of egalitarianism sets in, in which the team follows ideas and directions that feel right, regardless of who originated them.

The creative team seeks and rewards diversity. It actually celebrates and encourages ideational conflict in the context of providing individual safety for uniqueness. While individual differences will inevitably lead to some interpersonal conflict, the values of the team are such that facing and dealing with such conflict is not seen as overwhelming or destructive. It is seen as a normal result of different minds working together, related only to the incredible mental diversity that facilitates creativity in the first place. That is, the creative team relies on the diversity of its members to provide the synergy necessary for the breakthroughs and high levels of performance it seeks.

Performance feedback is given constantly, and in the spirit of individual and organizational growth, within a framework of trust. At the same time, rewards are shared in the creative team. After adjustment for individual levels of experience and expertise, compensation is shared equally among team members. Success and failure are accepted collectively. Responsibility for individual and team success is more widely shared; at the same time, each member holds himself or herself accountable for making the team a success.

The creative team is porous. People come and go from other teams and functions, both inside and outside the host organization. The creative team is connected to information and culture from around the globe. The team is open to information from every source. The team itself is most likely to be multicultural and ethnically, racially, and sexually diverse.

The creative team is constantly re-inventing itself. It continually revisits its vision, mission, goals, work procedures, and infrastructures, updating constantly everything from its tactical communication plans to its ongoing development strategies. Dedication to the values of the team remains constant even if vision and mission are redefined. These values reflect the nature of organic, living systems. They are the values that let individuals retain their uniqueness in the midst of a powerful work group. They reward invention and discovery, risk taking and creativity. These values require supporting others in a context of freedom and dignity, responsibility and fairness.

Figure 8-1 summarizes the team effectiveness continuum.

MANAGING TRANSITIONS

The idea that teams may exist in several states of effectiveness is not a new insight. On the other hand, most descriptions of teams as "high performance" or "dysfunctional" or some other categorization fail to examine their behavior at a depth necessary to understand how the team gets into and out of one or another state. The often-described team evolution along the continuum, "Form—Storm—Norm—Perform," is useful on a broad level, but doesn't tell us much about the issues the team has to manage within itself and its members to make these transitions happen. Even the efforts generally presented in the literature about teams and team behavior describe too broad a level to be of any help in identifying and managing specific behaviors.

A team's level of effectiveness is dictated by a combination of individual and team behaviors. Understanding the sophisticated ways in which individual behaviors interact in group settings to produce team behaviors is crucial to our ability to manage the transition to high-performance, creative teams. Individuals have to see and recognize their own and others' actions in the everyday context of the team at work before they can modify or enhance the behaviors that achieve the effectiveness level necessary for success.

To track these behaviors in a more visible way, we describe them as "dimensions" of team effectiveness. Dimensions are the characteristics and attributes that define the team's working relationships. Some very generic dimensions are present in every successful team, like working toward a common goal; other dimensions are unique to teams undertaking special tasks and challenges. When we ask groups to identify the dimensions required to make their own teams successful, a recurring set of generic dimensions, as well as those unique to the projects and team members involved, invariably surfaces. These dimensions can be captured and analyzed to see if they are holding us back or moving us forward along the team effectiveness continuum.

Each dimension mentioned can be tracked along the continuum. The state of the behavior within the team and as displayed by individuals will fall along the descriptors of the states as defined in Figure 8-2. Figure 8-3, therefore, shows how some common and fairly generic behaviors might be seen as descriptive of the state of team effectiveness. Another way of saying this is that the summary of the whole team's place along the team effectiveness continuum is reached by evaluating each behavior that the team displays, then placing it in the appropriate position on the continuum.

Managing the transition from one state to another is generally referred to by other authors as expressing control or discipline within the team. The individual, aware of the impact of his or her behavior, needs to check constantly for responses to situations and other members that are representative of the state the team wishes to achieve and maintain. Although every team need not always aspire to or constantly achieve the high-performance, creative

Chaotic	Formative	Functional	Productive	Creative
No defined or agreed upon standards for behavior and actions. The team is in chaos.	The state where standards for roles, behaviors, and actions must be defined to accomplish anything. The team must be directed.	Doing what needs to be done by following established rules, policies, and procedures. Taking direction from above. The team is externally managed.	Challenging systems and procedures—looking for new, more efficient, and better ways of doing things. Takes initiative by recommending changes to established rules, policies, and procedures. The team is shifting from externally led to internally led.	Inventing and creating new ideas, systems, procedures, products, or services. Takes initiative by making changes as needed. The team is self-directed, self-managed, and self-led.

FIGURE 8-2. States of Team Effectiveness

	STATE				
DIMENSION	**Chaotic**	**Formative**	**Functional**	**Productive**	**Creative**
Communication	Nonexistent.	Argumentative. Focused on defining values, vision, mission, and goals.	Smooth within the specific boundaries the team has set up for itself. Focused on getting the work done.	Lively and heated. Relevant information is shared within team. Focused on doing things better.	Free flowing and open. Unconditional information sharing, even outside the team. Focused on achieving team vision.

FIGURE 8-3. Communication as a Dimension of Team Effectiveness

state, we believe that future economic challenges, as defined by all the business and management literature of the last decade, call on teams to work toward that result. Figure 8-4 shows how some generic dimensions look in the creative state.

As we move along the continuum, behaviors change and so do results. One result of the basic behaviors at work in teams is trust. Trust is itself defined as a dimension by most teams, but is achieved when other dimensions are moving toward the creative area of the continuum. Trust, therefore, is based both on an attitude and on the results of other behaviors. Figure 8-5 shows how the dimension trust may evolve along the continuum, presumably as others support its development.

APPLYING THE TEAM EFFECTIVENESS CONTINUUM

Remember that the team effectiveness continuum is designed in concept and application as a descriptive and diagnostic tool, not as a rigid quantitative indicator. Therefore, its primary use is to prompt discussion and to illuminate the behavioral forces at work within a team so that members can see how the team might achieve its desired position along the continuum.

Some additional insights also apply:

1. Not all teams are required to achieve creative state. For some tasks and functions, another state may be most appropriate.
2. The team effectiveness continuum is not necessarily linear, and evolution does not necessarily proceed from chaotic to creative. A team may jump from one state to another in response to external or internal changes, or by decision or even command. In other words, a team at the productive state may slip overnight into chaos if the leader leaves or another person joins or the budget and scope change unexpectedly. The continuum is a fluid concept.
3. The definitions offered in this chapter result from years of observation and analysis, but teams may find that these definitions don't quite fit their situations. The definition of dimension, therefore, may differ slightly from team to team.

Dimensions have behavioral indicators. It is useful for each team interested in managing its way along the continuum first to

DIMENSION	STATE
	Creative
Communication	Listening to others, expressing opinions and thoughts openly and in a way that can be clearly understood.
Common vision	Holding a shared image of the future with members of the team.
Commitment	Dedication to others, to a cause, or to a work effort through difficult periods and easy times alike. An emotional and intellectual pledge of support and/or action related to a position or issue.
Conflict management	Facing openly and celebrating differences in attitudes, values, and beliefs. Working willingly toward achieving consensus.

FIGURE 8-4. Several Common Dimensions of Team Effectiveness in the Creative State

	STATE				
DIMENSION	**Chaotic**	**Formative**	**Functional**	**Productive**	**Creative**
Trust	Self only.	Trust your own preconceived ideas about others and the processes and procedures that you have seen work before.	Trust the processes and procedures set up by the team and only those people who adhere to them.	Trust clusters of individuals within the team to recommend and improve the processes and procedures. Trust the collective team.	Trust self and all others in team unconditionally to act in the best interest of the team. Trust both the team and the individuals.

FIGURE 8-5. Trust as Demonstrated in Each Effectiveness State

define its dimensions, then to describe the behavioral indicators that support that definition. In this way the team can assess what behaviors have to be changed, maintained, added, or eliminated to move it to its desired state.

Behavioral indicators are observable actions; individual team members and the team as a whole can actually be seen doing them. For example, look back at Figure 8-3. When we talk about communication as a dimension of team behavior, we could define it in terms of its behavioral indicators, the things people do when they communicate. A basic definition of communication by its behavioral indicators might be "The willingness to share feelings and information openly, the act of listening to others, and using available appropriate formats for doing both, including speech, written materials, media presentations, and computer-based technologies." We might resist such communication behaviors and wind up in the chaotic state, or we might excel at them and wind up in the creative state. Some basic definitions seem to evoke the creative state because, after all, they are ideal definitions and the creative state is the more ideal model of behavior.

Figure 8-6 offers examples of some other basic definitions of behaviors and their indicators. Remember that these are *sample* definitions; any team may alter them to fit their own culture and experience. We offer these from our experiences; as such they have value as starting points for your team's self-analysis.

Figure 8-7 is one possible format for such an analysis. By determining how the team performs against specific behavioral dimensions, the team can rate its overall performance on a behavior-by-behavior basis. Have team members behaved in a trusting manner toward most other members, most of the time? Or only toward closest associates? Does communication revolve around doing things better, getting things done, or what to do? When all relevant dimensions have been rated, assign values for each state in the continuum, from chaotic = 1 to creative = 5. Then determine the average state of the team by dividing the total score by the number of dimensions rated; in this case, 18 ÷ 2 = 2.25, the formative state. Members can then discuss how appropriate the formative state of effectiveness is for the job they do, which behaviors they want to develop, and at what state of effectiveness their team should operate.

DIMENSION	Definition	Behavioral indicators
Communication	Sharing information and feelings openly, listening to others, and using all available means for doing so.	Initiate interaction with others, share professional materials, offer to express opinions and feelings often, are expressive in meetings, and present ideas in professional formats continually. Ask others' opinions, and refrain from interrupting when others speak and present. Ask clarifying questions and support presentations with additional information, useful analysis, and honest feedback.
Common vision	A view of the future state or desired outcome shared by many.	Ability to articulate the same goal for the team that others articulate. Demonstrate understanding of the common vision through selecting which tasks are most relevant to accomplish at any given time. Demonstrate understanding of the common vision through behaviors that reveal attitudes in alignment with the vision, like achieving customer satisfaction before all else.
Commitment	A dedication to a course or direction, to a set of business goals, and to an overall vision for the organization.	Articulate the goals, mission, and overall direction of the business with enthusiasm and frequency. Display tenacity when work is tough and demanding. Display a sense of responsibility and ownership for the quality of the work, and for the well-being of each other. Demonstrate "stick with it" attitudes and work behaviors, including consistently sacrificing time and extending superior effort.

FIGURE 8-6. A Format for Analyzing Behavioral Indicators

Managers and team members need to be aware of certain essential steps in managing the flow from beginning to end of a team's life. For the sake of simplicity, we can identify the phases of the life cycle in the following way:

1. Start-up or project team launch
2. Team maintenance and development
3. Final assessment and team dissolution.

Start-Up or Project Team Launch

There is no more important moment in the life of a team than the start-up. What happens in the first few hours or days of a team's life will haunt it or support it for the rest of its work. As we identified earlier, teams are always born in some chaos. A carefully managed launch can do a great deal to move the team through the chaotic state and into more effective modes.

The best approach is to structure an off-site or special meeting where both social and work aspects can be addressed. Team leaders, team members, and those nonmembers who will directly support it should be present or involved in obvious and meaningful ways. An outside sponsor or champion of the team who does not take part in the team launch in some way may regret it later.

Whenever possible, the project team launch meeting should be facilitated by a professional who understands teams, teamwork, and team building. That professional should also be the acknowledged group facilitator to handle interpersonal conflict effectively and guide the group into agreement on the key issues.

A project team launch meeting needs to achieve distinct objectives, both social and work specific. The overall goals of the project launch meeting are to establish common objectives for the work to be accomplished and to agree on how the team will work together. Each objective is meant to cover the important areas teams must agree on to ensure both technical success and process success.

Process success refers to the way the team members interact to support the goals and work objectives of the team. At the conclusion of the project launch meeting, the team should not only understand the nature of the work and how it is to be accomplished, but also agree how it will work together to ensure success.

Eight important objectives of the project launch meeting, in more or less their most effective order, are discussed below.

☞ *Get to know each other.* Best done in a social setting like a company-paid lunch or dinner, this is where the off-site meeting idea really can work well. The evening before the meeting begins, members can attend a relaxing dinner followed by some casual introductory remarks from team leaders and sponsors.

The social gathering is not designed to make everyone best friends; that is not necessary. But it is useful to have some knowledge of other team members beyond the technical aspects of their work. Individuals who know something about each other's lives and interests and who experience the social sides of their teammates tend to understand each other better and, therefore, work together better. Knowing one's colleagues as people leads to a broader understanding of what motivates them and what they want from the experience, thereby helping the group bond around the common goals of its project.

Of course, there are always a few who resist the social connections, finding them distracting and unnecessary. It is better to let them have their way and not attend or attend reluctantly than either to force them or to cancel social functions for the others. In general, over time, even the resistant members will take part in social opportunities if the project team members and leaders as a whole build effective relationships, succeed at their tasks, and include everyone in the process.

Another aspect of the team's getting to know each other is not part of the social hour. This is built into the substance of the project team launch meeting and is more to the point of the work effort. Each member is asked to introduce herself or himself to the rest of the team and include several key points in the introduction, namely,

- Name and work affiliation, work description and history
- Skills and experiences brought to the team
- Expectations and aspirations for the team and his or her role in it
- Personal agendas to be addressed.

This last point, surfacing personal agendas, is crucial. Too many teams are sabotaged by individual, hidden agendas that are never

Another area of contradiction occurs as a result of our previous experiences with other teams. We tend to take our expectations and previous memories into the new team settings. Some of these can be helpful, but some can be hindrances to work attitude and output. As with the other areas of contradiction, the team must address its individual and collective memories to be sure they are managed positively for the team's benefit.

There are other contradictions as well, each of which must be investigated and resolved openly by the team as a whole.

☞ *Establish communication strategies.* The way we choose to communicate affects the quality of our interactions. Strategies are the overall approaches team members will apply to communicate among themselves and with outside associates. The team has to decide what kind of information will be shared internally and externally, by whom, how frequently, and in what formats. Controversial or high-visibility projects may require communication with management, sponsors, clients, and even the public much more frequently and with more control than a lower visibility, less controversial effort. Deciding these things with the best interest of all the stakeholders in mind is a necessary function of the launch meeting.

☞ *Determine project (task) management methodologies.* Deciding what systems and procedures will be used to track the technical aspects of the work is also on the agenda of the launch session. Agreeing on work processes and infrastructures, and being sure they are implemented are the specific tasks.

Work processes are the everyday procedures that drive a project forward. These are facility use schedules, test schedules and procedures, even expense forms and weekly reporting systems. All the day-by-day procedural activities and their documentation and support tools constitute work process considerations.

Infrastructures are the overarching technologies and policies that support the daily work. If a work process is to report test results daily, then the infrastructure is the computerized support of that reporting, including telephone systems and facilities. The infrastructure is also the organization of the team or the company within which the daily work is accomplished. An organization chart represents an infrastructure designed to enhance the daily work of the individuals within it.

The project team, then, must define, secure, and implement its own infrastructures and work processes to support its overall mission.

☞ *Establish operating principles and ground rules.* Operating principles drive the team's interpersonal and work relationships. Protecting the self-esteem of other team members is an operating principle. Sharing all information, respecting the privacy of others, and valuing individual differences are principles. These translate into ground rules, which are the guidelines for day-by-day behaviors in the work setting. Coming to meetings on time, not interrupting others, supporting each other in times of stress and pressure, getting reports and deliverables done on time, answering phone calls promptly, even cleaning up the coffee mugs are all day-by-day ground rules for making working together more enjoyable and efficient.

Team Maintenance and Development

During the time a project team is working, constant monitoring of the work process and interactions is advisable. This means checking with the group to be sure that initial decisions are being followed. A process check should be part of all team meetings. During the process check, members are encouraged to surface any concerns or issues they have about how the team is working together.

One effective way to manage a process check is to review the initial agreements made at the project team launch meeting. Sometimes aspects of those agreements will need to change. Perhaps someone's role shifts, or maybe some agreed upon ground rule isn't working and must be renegotiated. More seriously, conflict may have erupted between two or more members and must be addressed before it affects the well-being and effectiveness of the entire effort. The process check is the place for these issues to be raised and managed. Occasionally process issues may arise that are outside the skills of the team leader or members to deal with alone. The human resource department or an outside team consultant may be needed. The momentary relief of avoiding process problems is not worth having the entire team seriously set back later by unresolved conflicts and misunderstandings.

Along the way, the team will probably want to strengthen its skills and capabilities, both technically and as a work group. The team effectiveness continuum can help here, too. Use it to assess the position of the team along the spectrum from chaos to creativity, and examine which of the dimensions outlined earlier are working well or in need of attention. For example, the team may assess itself as working smoothly in the productive mode, but notice that certain aspects of interpersonal communication occasionally slip into chaos. Determining why that is happening and addressing the reason is essential. When the problem or source of the trouble is identified, the team can address it through renegotiating role assignments or work relationships, holding a special skills training or problem-solving session, or using some other technique that will drive communication back to the functional or productive level.

Also important are the more traditional aspects of team management, such as monitoring work flow and quality, tracking timelines and budget milestones, and monitoring technical progress against expected specifications and final results. These also must be reviewed continuously, using traditional project management tools, so there are no surprises for technical sponsors, customers, or internal management.

Final Assessment and Team Dissolution

The last phase of the formal life cycle of a project team is its dissolution. It might be tempting to think that when the last deliverable is completed, we shake hands one Friday afternoon and come in Monday morning to the next assignment. All too often that is just what happens. In ending project teams that way, we miss an important opportunity.

Looking back over the project with an evaluative eye, the team should revisit the successes and failures of the effort. This should occur in a meeting just like the launch meeting, with no other agenda but to capture the experiences of the team members.

Each of the areas discussed at the beginning of the project (listed on the previous six pages) should be reviewed, both process issues and technical or project management issues. Then with the help of a facilitator, these comments should be prepared as a final report about the project itself and made available to other project teams.

In addition to this formal evaluation process, there is always room for celebration and reward. Teams that have worked hard together often forget to thank themselves and celebrate success. A dinner out together with spouses, a few special awards for various kinds of contributions, and in extreme cases, perhaps even team bonuses or trips might be appropriate. Whatever the level of celebration and reward, some acknowledgement and pause to thank makes individuals feel important to the process and reinforces the best behaviors of teamwork itself. In addition, it may be an important contribution to the mindset with which the individual members of the team approach their next projects.

SUMMARY

Some kind of organizing principles are needed to understand, participate in, and manage teams. The team effectiveness continuum provides such an organizing principle on a behavioral basis, meaning that team members and leaders can look at the cumulative effect of individual attitudes, beliefs, and behaviors on the overall performance of the team. Furthermore, using the continuum as a diagnostic tool allows teams to track their development and adjust their skills and behaviors to shift their performance level to whatever position along the continuum makes sense for the work at hand.

Teams are a crucial part of the business and technical world today, and they will continue to grow in importance as the world becomes more technologically, economically, and socially interdependent. A crucial survival skill for both individuals and organizations will be the ability to work efficiently in a team. This means personal and group level agility and flexibility as we move from project to project and team to team. Note the irony: the success of an individual's career will depend on his or her ability to perform uniquely and creatively in a team. Set aside the old Horatio Alger myth of "going it alone." The well-being of our organizations, our businesses, and our selves is now a team issue. We need to replace the heroism of the Lone Ranger with the heroism of the team. As Robert Reich said, we need to tell ourselves the stories of teams at work and teams succeeding until we learn how to do it and do it well:

> Stories about Triumphant Teams—composed, like any assemblage of Americans, not of selfless, subservient drones, but of creative, idiosyncratic individuals, yet devoted to common goals and committed to reaching them through common efforts—may well already be working their way into our mythology. Triumphant individuals are being replaced by collective entrepreneurs.*

Everything hangs on our common effort to succeed uncommonly.

---

*Robert Reich, *Tales of a New America.* New York: Times Books, Random House, 1987, p. 246.

CHAPTER 9

MANAGING WITHOUT AUTHORITY

A NEW DEFINITION OF MANAGEMENT

Much is said about delegating power to others, spreading out responsibility for projects and work requirements, and generally involving the work force more and more in decision making and accountability. This is great advice for bosses and upper management, but what if you are on the receiving end? Just being told to take charge and get things done may not be enough. After all, many of the people whose help you need to succeed may not even work for you. Sometimes responsibility is delegated to a level of staff members who have no one reporting to them. How can you succeed at managing with little or no position power to draw on?

In the past, almost all organizations structured power through the hierarchy so that everyone would understand the lines of authority and responsibility. However, like so many other areas of corporate and organizational life, this too is changing. In most organizations, the nature of work has changed drastically since the

- The need to control others—*Domination*
- The need to build work and social networks—*Affiliation.*

All three carry positive and negative aspects. The need to achieve could be neurotic. Consider the classic, out-of-control, "Type A" workaholic, who is ready to take herself and everyone else with her down the path of stress and burn-out just to look like something is being accomplished. Her need to control others is motivated by a need for personal glory and benefit. On the other hand, nothing gets done in the world unless someone is motivated to do it, to work with diligence and intelligence toward a goal that she believes in. Involving others in that process is a matter of necessity and pride, not manipulation and usury.

The need to affiliate can manifest in building empires and flashing around personal ego. The more noble side of affiliation is teamwork and team building. Affiliating with others can mean building a work community in which people share success by sharing skills and expertise in a way that leads to creative output and mutual on-the-job satisfaction.

Of the three forces that motivate people to aspire to positions of power, the need for achievement dominates. Getting work done through others means using power as a positive influence. Position power clearly enhances and facilitates the ability to get things done through others. *The irony is that the need to accomplish work through others in the modern organization is even more acute, and the availability of positional power even more scarce.*

Many of the best influence skills are good interpersonal skills, thoughtfully applied. Some key attributes used successfully by influence managers are discussed below.

Knowledge of Others

The successful influence manager understands the motivations and needs of others. Appealing successfully to those needs means he gets commitment from others to help, to give time and expertise, and to see his needs as a priority in their own work. Motivations, of course, vary from person to person. For some the mere opportunity to work on an interesting technical challenge is enough; for others, who else is on the team may be a great motivator. Some might be motivated by the visibility a project offers;

others might see in it the opportunity for acknowledgement of their technical expertise. Most people will not voluntarily buy into an influence manager's request for help unless they know what's in it for them. Finding that out and appealing to the people he needs on their terms is the strategy here. Translating the goals of the request for help into terms that answer "What's in it for me?" is crucial.

Communicating Effectively

To be successful at any of the strategies defined above, the influence manager needs exceptional communication skills. The first of these is the ability to listen. Not just hearing what others are saying, but listening to what they are saying is the definition of this attribute. When someone indicates a "What's in it for me?" motivator, can the listening influence manager recognize it? Can she pick up on it and build her case around it? The usual approach we take to hearing is to let others finish, then say what we were going to say anyway. The listening manger uses the information provided by the person she is communicating with to build her case.

In so doing she makes her own commitments as well. If she listens and hears the person say he needs to be recognized for his technical expertise, and she says that will happen, she had better make sure it happens. Listening means more than just manipulating the information offered for immediate advantage, it means remembering what others have said and following through on it later. In short, proof of the quality of listening is in the subsequent behavior of both parties. The influence manager does a lot for herself by listening well, committing to deliver what she has promised in return for the other person's involvement, then following through on the deliverable. The real result of this is that word gets around that this influence manager is trustworthy. It's much easier to get help in the future if your past is a credential, not a liability. Listening and following through is one of those credentials that will always work to your advantage.

Facilitating Conflict and Assisting in Difficult Situations

Nearly as important as communication skills is the ability to help others handle difficult situations. Usually these are conflicts between work colleagues. Influence managers possess conflict resolution and management skills because they have to, to keep work groups on track. Recognizing interpersonal conflicts early and being available to help with them before they get out of hand and undermine the work is crucial. When people trust that a team leader will come to their aid with a stressful interaction, they are relieved and grateful. In addition, early and effective conflict management keeps projects on track and allows the conflicting individuals or groups a way to work together even though they may have different ways of perceiving things. The successful influence manager rises to these occasions, he does not bury the problems or avoid them. In this way, he strengthens his reputation as a successful leader because word gets around that he can be trusted to keep people focused and on track, dealing with the hard issues efficiently along the way. Sometimes he does this by managing negotiations, by mediating, by helping individuals or groups collectively solve problems, or by carefully counseling individuals to adjust irritating behaviors or habits. Whatever approach he uses, it will be effective because he will use his communication skills, listening and following through, to help manage conflict and other difficult interactions.

Building Strong Interpersonal Relationships and Networks of Support

The successful influence manager knows she can't do it by herself. That's why she is able to ask others for help without feeling inadequate. She gives credit to others, appreciates them for who they are, and spends time caring for her relationships with them. She thanks people, remembers their important celebrations, and makes herself available to help them whenever possible. In short, she nurtures her relationships and in so doing maintains an active network of business friends and colleagues.

She is building investment accounts. She remembers to deposit something of value into another's account when she asks them to give her something of value from that account. If I ask you to cover on a presentation for me this week, you are more likely to agree if I have a history of giving meaningful favors and supportive behavior. Sounds like banking IOUs, but it's more than that. The value added of these deposits and withdrawals from each other's accounts is the knowledge that we are protecting our relationship and our business trust by watching out for each other's needs. At best, it's an expression of interpersonal trust and respect, and even at worst, it's a useful arrangement for both parties.

The result of these investments is an active network of support for almost any need the influence manager might have. She has developed a list of diverse and exceptional talent and a willing supply of professional friends to help out.

Personal Flexibility

The successful influence manager is a model of agility. He knows that to work effectively with many people, he must not be seen as rigid and difficult to work with. He can adapt his style to others' needs to help them feel comfortable about working with him. This does not mean that he lowers standards or lets project timelines slip. On the contrary, he is firm about both quality and timeliness, but aware of others' work styles and strategies. Therefore, he knows when to accommodate, when to support, and when to push back, but in the context of situational and personal uniqueness.

Another way he shows his agility is by being flexible about his own work style and timelines wherever possible. Since he is asking people to work for him without the authority to command them, he must be more sensitive to adapting his schedule to theirs, going out of his way to find the support someone may need, and generally displaying the best model he can of openness to the various ways individuals work.

Demonstrating Positive Energy and Vision

The successful influence manager demonstrates charisma in inspiring others to reach goals and work to build a future. Clearly articulating the goals and the desired outcomes is more than assigning work. It means turning people on to the work and the potential success that comes with it. Even at times of distress and setback, the influence leader continues to bring out the best in people and maintain focus on the effort ahead. She demonstrates a positive energy that is contagious and keeps people going no matter what the obstacles.

The key to this ability is actually an emotional aspect. The successful influencer has to feel a stake in the work she is doing. Faking that won't work. To get people to work for you on the kinds of efforts influence leaders usually have requires unmistakable enthusiasm and a genuine belief in the vision itself. No one will be fooled by a leader who doesn't believe in what she is asking you to invest part of your life. The successful influence manager chooses her "cause" carefully (whenever possible) to be sure that her heart is in it. She can't expect others to put their hearts where hers isn't. After all, she is managing from the authority of her own heart more than anyone else in the management arena.

Personal, Business, and Technical Credibility

In addition to being goal focused, the influence manager has a reputation for getting things done. With his technical and business expertise well known, others trust his ability to perform. Building a track record is the key here. People always want to be associated with a winner, and the winning influence manager is an attractive option with whom to align. When pulling teams together, the persuasion of note is the past accomplishments of the manager. This is as it should be, since each of us would like to enhance our own career and reputation by being involved with people known as movers and shakers.

Most important, though, is his personal credibility. He not only must believe in the work, he must be known as honest and forthright. He must be seen by others as someone with whom they want to be affiliated because of that integrity and because he can be trusted with their careers and aspirations. This trust based on personal credibility may do more than any other single thing to attract people to work with him and to keep them coming back when the opportunities are presented.

Building and Managing Effective Teams

Since teamwork is such an important ingredient in today's work environment, the successful influence manager spends a great deal of time and effort on this aspect of her leadership. Helping groups work together, even though their backgrounds and motivators might be quite different, is the core of this effort. Modeling teamwork by demonstrating, wherever possible, a participative management style helps keep the team intact and working well.

Actually, building teams and fostering team work is an exceptionally good technique for strengthening personal authority, since teams basically enjoy and need a leader to guide them through the maze of tasks and politics that surround most projects. In a team setting, there is less need for a "boss" in the classic sense than for someone who is generally successful because of all the traits discussed here. As leader of a team that works well as a team, the influence manager finds the best environment for her particular brand of management. Finally, some of the best results having the most impact in the organization can come from tightly bonded teams accomplishing specific objectives in a short time.

The successful team is the best place for the influence manager herself to get further visibility, credibility, and respect among other leaders in the organization. After all, even influence managers may want to have positional power someday.

SUMMARY

Influence management is an important and evolving skill in today's work environment. Here we have introduced some of its basic tenants. Of all the capabilities worth developing consciously over the next decade, positive power and positive influence are two core skills not to be ignored. The future will belong to the managers and not-yet managers who can focus the broadest array of resources on the complex and demanding tasks of the competitive organization.

heightened awareness of the value of this potential as well as the development of specific new behaviors. In this chapter we will discuss some of these in detail.

What is meant by "creativity" and "innovation"? Creativity is most often used to describe individuals and organizations who have discovered or invented new solutions to major problems or concerns in a field of research or product development. Innovation is generally used to describe the gradual process of improvement or enhancement to existing ideas or products through continued tinkering and trial-and-error refinement. For example, the Japanese have been considered "innovators" because they have refined "creative" ideas and inventions from American technology into highly effective electronic products.

Two kinds of creativity seem to recur in normal conversation. Capital "C" Creativity seems to be the Creativity displayed by Mozart, O'Keefe, Einstein, and Hawkings. This kind seems genetically programmed to produce super-genius discoveries and thinking mechanisms: brains untouchable by most of us, so fantastic are their creative properties that they give the creative process a kind of hallowed distinction. An attitude of awe surrounds the names of these famous geniuses, and they evoke in us a sense of wonder. Few of us aspire or imagine ourselves to reflect in our own lives and work the talent of such geniuses, and we recognize their giant stature among us.

Occasionally, a manager may be called on to manage an individual with this kind of capital "C" Creativity. This presents special challenges since scientists and engineers, like poets and musicians, often display unique and difficult personal characteristics at this level of creative genius. As a manager of technical professionals once said to me, "The truly Creative people will always do it their own way, and all the manager can do is keep the resources available and other people out of their way." Some might argue that is a great definition of management in general. Somehow the nearly eccentric behavior of the truly Creative people pushes the edges of almost everyone they come in contact with.

Little "c" creativity is the kind that is expressed by many of us in day-by-day ways. The creative snapshot we take of a friend, the creative way we decorate our home, office, and even ourselves when we dress in the morning. We may plan creative vacations or choose a creative gift. At work, little "c" creativity shows up as a well-constructed report, insightful conclusions drawn from rigorous analysis, a breakthrough application of a new technology, and so forth. We often perceive this creativity as merely good work and don't recognize it as creativity at all. In fact, in hundreds of interviews with people who perform all of the acts just described, people tell us they are "not creative at all," or they "can't remember the last time they did something creative." They seem to be referring to capital "C" Creativity and are not at all cognizant of the ways we are creative all the time.

If managers could help their teams and individual contributors recognize the value of little "c" creativity, perhaps the characteristic of personal creativity would be more widely accepted. By raising awareness of the creative things people do and are capable of doing on a regular basis, whether entry level technician or senior scientist, we could build an environment of enhanced respect of self and of the creative capabilities we display daily.

How can managers acknowledge, reinforce, and foster the creative aspects of the people and teams they oversee? What characteristics should they reward and encourage? Most importantly, what kind of creative thinking and actions should they model for others to emulate?

Some characteristics of creativity and creative people are in all of us. The notion that creativity is available only to a select few is naive and outmoded. We all have the ability to create and, more important, the ability to learn to be creative. Even little "c" creativity can be pretty powerful stuff and make a huge difference in individual and group success.

However, we need to learn how to encourage creative thinking in ourselves and others so we can develop a truly viable skill set. A *skill set* is a collection of attitudes and behaviors we can employ at will to accomplish tasks and get measurable results. Creativity offers two categories of skill sets. The personal skill set describes our individual abilities to harness the creative process and make it work for us, and the skill set the manager draws on elicits creativity from others.

Looking more closely at the first, we can see that personal creativity, when left to itself, will occur spontaneously and with results. But creativity understood and managed will yield its conscious structure and manageable elements, which can be enhanced and developed as a tool for productivity.

The sister of creativity is innovation. If creativity is the invention and discovery of new ways of seeing things, then innovation is the slow refinement of ideas, products, services, whatever. In fact, in the business world over the last decade, the Total Quality Management movement actually represents a kind of structured innovation process. TQM fosters a step-by-step quality analysis process integrated throughout the whole production cycle of any organization's work; it functions as a continuous improvement mechanism. This doesn't replace creativity; it augments and builds on it.

THE CREATIVE PROCESS—AN OVERVIEW

What is creativity? Some have called it the ability to see things differently. Others, the ability to break out of ruts by challenging our thinking patterns. Still others have called it purposeful play. Some see it as connection with some universal source of knowledge, insight, and new ideas. Others equate creativity with imagination and urge us to develop our powers of vision and intuition. Einstein said that imagination was more important than knowledge. Still others have called creative only those inventions that solve practical problems and can be proven useful.

Whatever others say about it, one thing is certain. Only the definition of creative that you construct for yourself really matters. You as an individual and your work group will carve out your own definition and application of creative thinking and behavior, those best suited to your own life and work style.

Is there a "creative process"? Literature is filled with research about human creativity and how it happens. Many see creativity as a human experience, a recurring mental process that is observable and documentable. Although the process may vary from situation to situation and person to person, some of its characteristics are consistent. A standard view of the creative process and how it unfolds has been refined over the decades of the 20th century. We offer it here for your information, as a very generic template for consideration. The flow of the creative process as discussed in volumes of literature looks like this:

☞ Initial Interest

☞ Preparation

☞ Incubation

☞ Illumination

☞ Verification

☞ Implementation

☞ Evaluation

These phases capture the assumed flow of thinking we put ourselves through when searching for creative solutions or generating new opportunities in the face of challenge. They can be expanded upon and better understood within the more broadly defined and manageable context represented in Figure 10-1. Among other things, this cyclical view of creativity offers a more structured approach to managing creativity in ourselves and others, especially where bold new solutions are needed for technical and business applications.

It is clear in Figure 10-1 that each phase of the creative process is complex. It may be less obvious that the cycle itself is iterative, not even circular, and never linear. Each phase consists of a number of specific activities that we can consciously manage. These are represented along the outer circle of the figure. Taken as a whole, they constitute the boundaries of a highly manageable and structured approach to the creative process.

STRUCTURED APPROACHES TO ENHANCING CREATIVITY

There are as many models and theories of creativity and creative thinking as you have time to read about—"whole brain" models, business success models, social style inventories, and analyses you can fill out until the end of the century. What they all boil down to is this: How do you tap into your own creative potential and make it work for you in achieving personal fulfillment and life success?

To achieve anything, we have to act. Essentially, creativity is novel ideas that find their way into action. Ideas that stay ideas go nowhere.

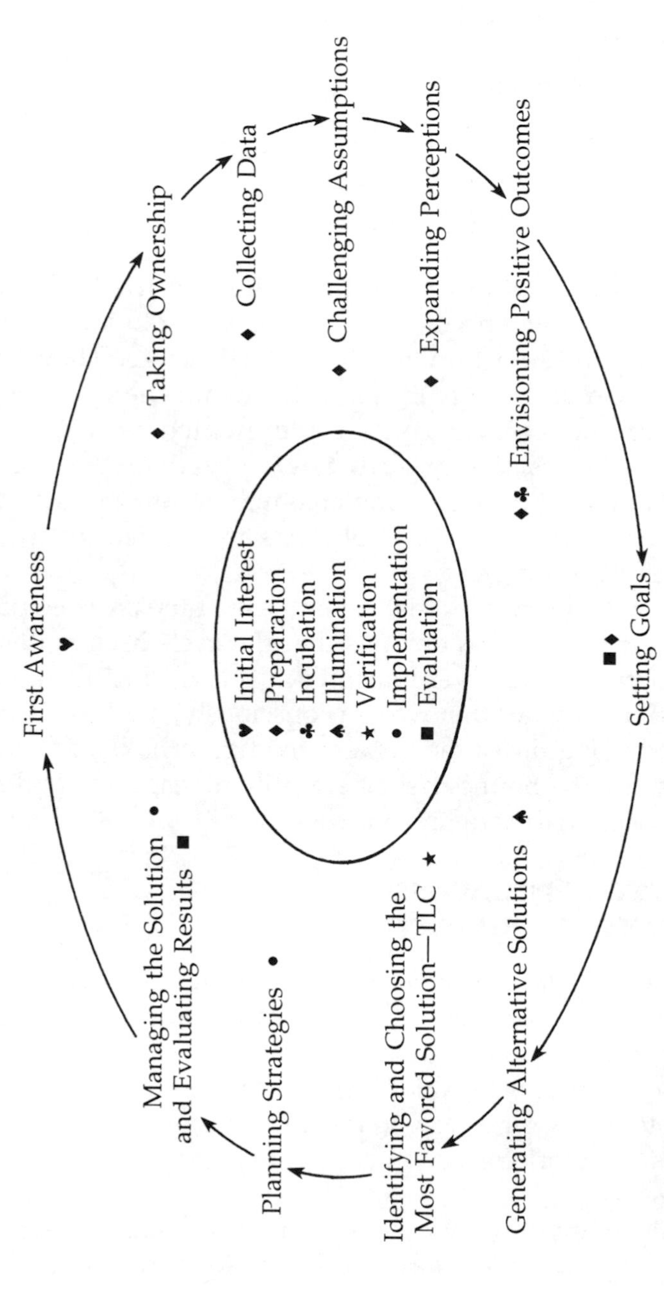

FIGURE 10-1. Creative Thinking

Nevertheless, the starting point for creativity is in the way we manage our minds. In fact, as we travel around the world working with individuals, teams, and organizations to enhance creative thinking and actions, the line I hear the most is, "Oh, I'm not a very creative person." Where on earth did we ever get such notions? Unfortunately for most of us, and for our children, we were told this at an early age.

The perception of oneself as not creative is what keeps people from discovering and using their innate creative capabilities. Creativity is the human condition: "All human beings are capable of being creative—it is part of our birthright."* The end result of the structure and organization of the brain and, therefore, consciousness is by definition creative. We are able to sense and respond to the world in ways that no other animal can because we have built in the hardware for recognition and manipulation. This sophisticated and complex system is made up of the five senses (that we know of), which constantly collect data from inside and outside our bodies, and the brain, which is constantly processing their input.

The key to creativity is raising our own awareness of how our senses and our brain work together. This allows us to see and appreciate our own brand of thinking, acting, and creating. Only then can we begin to manage the process and enhance our own ability to sense and process the world around us.

Creativity can be structured into habits and beliefs like most other personal and work skills we develop. Developing the discipline for creativity is not necessarily easy. Some of the most disciplined forms of creativity, like painting, music, and the other arts, are in fact extremely challenging disciplines—so challenging that they are difficult to teach and learn, which is one reason they are so sorely neglected in Western education. The so-called "soft" studies are actually among the most disciplined of human behaviors. If you don't believe me, try living with a musician and watching the effort behind what looks so easy.

---

*Herrmann, Ned, *The Creative Brain.* Lake Lure NC: Ned Herrmann/Brain Books, 1990, p. 184.

So "structured creativity" is not the self-contradictory phrase it would seem to be. No filmmaker ever made a great film without enormous discipline, from managing time and budgets to visualizing, shooting, and editing the action into a coherent experience for the audience. No artist ever painted a painting without adhering to the rules of color, light, and form while creating something totally unique.

This kind of capital "C" Creativity—the kind that we think of when we consider people like Mozart, O'Keefe, and Einstein—only exaggerates the point. Since we are all creative, albeit little "c" creative by and large, we also have need of discipline and structure in approaching creative challenges. Little "c" creativity is by no means small potatoes. It allows us to produce products and services, relate to others from customers to teenagers, build and decorate our homes, and invent our lives.

We need discipline to enhance creativity. Our starting point is to recognize that creative thinking and behaving have definable elements. Very often these elements unfold in a pattern, sometimes even in a sequence. Starting with the recognition of an initial interest, challenge, or problem, we consciously trigger our senses to gather data, although unconscious data collection was likely the source of our initial awareness that the challenge exists.

Initial Interest

Our interest in or awareness of an issue can come from many sources. Perhaps a technical problem surfaces that we had not anticipated, or an interpersonal struggle of some sort materializes in the work team. Sensitivity to problems and challenges differs from one person to another. Some of us anticipate interpersonal conflict more easily than we recognize the clues of an impending technical disaster. Others easily see scheduling problems as they surface in a project management situation, but never anticipate a coming financial crisis.

Why some of us are more or less sensitive to a particular kind of problem is probably a result of certain dominant thinking preferences we have developed. Those sensitive to people and their needs see people problems easily and usually don't hesitate to confront them in a timely way, whereas those sensitive to technical matters can see the clues of technical trouble a mile away. It may be possible to train ourselves to be more sensitive to a variety of issues earlier, and that would be helpful for managers who are supposed to be on top of the complex environment they manage. At first, though, we can recognize and affirm the areas we are naturally sensitive to, and focus our problem-solving efforts more rigorously there.

Early recognition of problems is like early diagnosis of a physical ailment. The faster we get on it, the more chance we have to intervene before the problem damages our business or technical work. Delayed response to challenges can retard our efforts to find creative solutions. The increased stress associated with coming at a problem that has already developed may compromise the quality of our creative process and, therefore, our solutions.

It is surprising how advanced some business and technical challenges become before they are accurately diagnosed and a concerted effort to solve them emerges. In the technical world, this seems to be especially true of difficult interpersonal communication and relationship issues. The avoidance technical people show to people problems causes these problems to be recognized later in their development, thereby adding to the difficulty of managing them. Letting people problems fester, of course, results in diminished productivity. Since all of the ingredients in an organization require our attention, managers would do well to become more interested, knowledgeable, and sensitive to a wider range of organizational and institutional issues. Only by anticipating challenges and problems can we turn them into creative opportunities. When they become crises, we respond with bandaids, not leadership.

Preparation

Louis Pasteur said, "Chance favors the prepared mind." The more we work with an issue, the more saturated we become with information about it, the more likely the brain will make connections and see possible answers. But preparation is not only a matter of gathering facts and figures; it is also apparently experiential, even emotional. Caring about the challenge, interacting with others about it, and experiencing its affects are all visceral elements that further motivate and involve our thinking mechanism in the search for answers. Einstein said, "I hardly ever think in words." He was expressing his visual and mathematical approach to the universe. Einstein experienced major problems of cosmology as visual metaphors, which for him translated to mathematical models and equations. His preparation step (which lasted much of his life) was full of visual imagery (daydreaming), numerical manipulations, and conceptual play.

Purposeful play may be the key to preparation: looking at things differently, expanding our view of everything connected to the problem. Preparation has been synonymous with gathering information, but the preparation step is richer, deeper, more totally involving. By studying the behaviors and components of preparation as an aspect of creativity, we can better understand Pasteur's point.

Taking Ownership of Problems and Challenges. As we become aware of the issue, we must embrace it with all our energy. Taking ownership means focusing energy on the problem and abandoning any sense of victimization we may consciously or unconsciously be harboring. The most effective way to take ownership is to restate the challenge in action words, with specificity and clarity. Translating our language and our thinking from "The problem is..." to "How to..." words helps us take that ownership. "How to..." thinking is self-igniting and self-motivating. It is active, rather than passive. "How to..." thinking redefines pressure into potential and allows for action. So the first step in creative empowerment is to translate problem statements into achievement statements.

Create as many "How to..." statements as possible for the issue until you can't think of any more. Using well-known brain-storming procedures to generate alternative statements will help flush out details and move you closer to the core of the challenge. The rules to follow when brainstorming are to suspend judgement, go for volume of ideas, capture everything, be wild and crazy, and build on each others' ideas and comments. It is one of the most useful techniques we have for moving toward a more accurate definition of what the problem really is.

Collecting Data. As we gather data about the issue, we shift our view and understanding of what is going on and what the issues are. The data comes from anywhere we care to look. Reading, observing, asking others, and searching our own minds and experiences for what we already know and feel about it. Data can be collected in many formats. Perhaps we collect articles and amass books on the subject, find TV programs and videos to watch, or visit places and people involved with the topic. We may even use our own personal experiences and the hunches we have gathered over time. Any and all sources are welcome at this point. We ravenously gather facts and impressions from wherever they come as a way of expanding our perceptions of what the issues are and what the challenge really is.

Challenging Assumptions. We live our lives by assumptions. This or that is the case, such and such will happen if.... In the creative process, all assumptions are open to investigation. During the 20th century scientists have been forced to challenge nearly every assumption on which our understanding of the universe is based. In so doing they have re-invented reality!

We need to be willing to challenge our assumptions. How do we know what's happening here? Is our source reliable? If the opposite of the operative assumptions were really true, then what? Challenge the linguistic and hidden meaning behind every word in our "How to..." statements. Ask uncomfortable questions about the assumptions of others, and question our own experiences to probe for deeper insight. The road to hell may really be paved with unchallenged assumptions!

Expanding Perceptions. Human perception is a marvelous and complex mechanism. We collect data from our five (known) senses and ask our multiprocessing brains to interpret the input. We expect that intelligent insights, workable solutions, and effective courses of action will result. Whether they do depends on how well we manage the perceptions themselves. Our senses, after all, play tricks on our brains literally all the time. For example, have you ever approached a familiar car in a parking lot and inserted your key in the door lock only to feel it refuse to turn? A wave of disbelief and confusion comes over you as you ask, "What's wrong with this lock?" Suddenly your shocked perceptual system has to reassess what's happening, and you realize that it isn't your car at all. An imposter! A look-alike parked right where (you thought) you parked yours. Only when you become able to recognize that something is going on other than what your original perceptions indicated can you begin to get a handle on the real situation. Embarrassment and confusion may overtake you as you back away from the scene to look again for your car. And how you hope no one has noticed!

Perceptions, like assumptions, must be constantly challenged. We have to train ourselves to back away and look at things differently.

For example, after reviewing the monthly financial data, we might decide that we won't make enough money this quarter. That may be the case, but what else could be going on? Perhaps we are spending too much, or maybe we are making necessary investments in the future that give the illusion of red ink.

The processes of challenging assumptions and expanding perceptions work hand in hand to help us investigate the true definition and characteristics of the problems and challenges we face. Just asking the question, "What else could be going on here?" will keep our perceptions fresher and our awareness of the problems and challenges before us more acute.

Envisioning Positive Outcomes. If we can imagine a future state, we can invent it. That's what Alan Kay said.* The future state is that condition we hope to achieve by solving our problem or rising to the challenge. It is not necessary to know what the solutions are to visualize the future condition, it is only necessary to engage the imagination and release ourselves from the obstacles of the moment. We can do this in several ways. First, we can create a guided imagery in which we are transported by suggestion to another time and place. This usually requires a quiet room, some pleasant music, and a relaxed frame of mind. We can do this alone or with others; we can be taken through it by a live facilitator or a prepared audio tape. After the session ideas and visions are captured on paper or discussed in groups for retention and development.

Another approach is to work with one other person or in teams to brainstorm the ideas that come to mind when we think of a

---

*Alan Kay is a computer genius who specializes in artificial intelligence. He is involved with MIT's Media Lab, where he heads the Vavarium Project. More on this is described in Stewart Brand's book, *The Media Lab: Inventing the Future at MIT* (New York: Viking/Penguin, 1987).

future with the problem solved or the challenge met. Ideas and insights generated by the group are charted and saved. The picture that emerges from such group work can form the basis of a corporate or departmental vision statement.

Setting Goals. Goals derive from the vision. They are the specific measures that will tell us if we have met the challenge. If visions are right-brained, then goals are probably left-brained. They are measurements, and they give meaning to our achievements. If the vision is of a technical environment where scientists are working productively and are happy at what they're doing, then the goals will prove that. Productivity will increase by, say, 18 percent and employee turnover will drop by 6 percent. These are measures that might be used to prove that the vision has come to fruition. Setting goals helps us prepare for finding solutions by bringing a level of specificity to the search. They lay the groundwork for other steps in the process, like implementation strategies and evaluation criteria.

Incubation

This step grows out of the preparation process. When, like a wet sponge, we are saturated with information, feelings, analysis, interactions, possibilities, and need to get away from the problem, we incubate. Forget about the problem, leave it for a while. Take a walk in the woods, play a round of golf, or spend a week on the beach. More likely, though, we work on something else. Meanwhile, like a cake mixed and placed in a hot oven to cook, the ideas take shape. Like a background computer program, the brain churns away on the problem effortlessly (it seems to us), weighing all the ingredients and trying out new arrangements of the information we have provided. Incubation can take a few seconds or many decades. A person's entire creative life can be measured in iterations between preparation and incubation, waiting for that next important step, illumination.

Illumination

Generating Alternative Solutions. Well-stated problems or challenges help the search for solutions. A less creative choice than searching for new solutions is to choose from existing solutions one

can remain a part of the creative process. For example, since so many new ideas are questionable at best and often strike us as crazy, we immediately conjure up all the reasons why they can never work. Here are some of them:

> "We tried it before and no one liked it."
> "The boss will throw us out of her office."
> "It costs far more than the budget can handle."
> "It's impractical."
> "We better let the committee look into it."
> "It's never been done before."

And after a day in any mainstream American company, I could add twenty more "good" reasons why new ideas won't work. The landscape is full of devil's advocates who play why-it-won't-work games with every new idea that comes along to threatened the *status quo.*

Yet we need to evaluate ideas, so how can we do that without killing them?

One answer is to let the verification step be a part of idea development and evolution. This can be done with a remarkably simple tool, tender loving care. All we do is modify our devil's advocate behavior enough to make it angel's advocate behavior. The angel's advocate wishes to save and evolve the idea, no matter how shallow or impractical it may be in its embryonic state; the devil's advocate is out to destroy it, no matter what words he or she chooses. In the creative environment, the question, "May I play devil's advocate?" is answered firmly with, "No, we need an angel's advocate. Tell us how we can provide tender loving care, instead."

Translate tender loving care into three questions:

1. What is tantalizing about this idea? What makes it interesting and worth pursuing? What is good about it?
2. What is limiting about the idea? What points does it miss? What other potential problems or liabilities does it suggest? What's wrong with it?
3. What could be changed to make it the perfect solution?

Now we are verifying—testing, questioning, evolving the idea into something really quite useful. When we have honed it through this process, we can apply further rigorous testing to see if it can enhance financial gain, meet operational standards, integrate into the existing organizational culture, etc. But we must always be willing, as we will explore later, to let the idea have priority, give it shape, and challenge ourselves to embrace the new, not for its own sake, but for the advantage the right new idea will give us in the long run.

Implementation

Planning Strategies. This is the step we are best equipped to meet. Planning to implement the new idea requires all of our best tactical skills. We assess current systems and determine what will have to change, we plan the people and technical resources necessary, then we plan what steps must be taken when and by whom. It's project management at its best. We construct budgets and operational guidelines for a new manufacturing technique, an organizational change, or a modified research approach. Whatever the idea demands of us, we translate into actions that are monitored and managed until the new idea is implemented.

Managing the Solution or Innovation. Now we have come full circle. The creative solution is part of mainstream work. Our process has changed the way we work and accomplish daily tasks. A new product, process, or service has become a reality. We are changed by it. Because one part of the equation has changed, everything is affected. It is through these creative processes that things are different today than last year, last decade, last century. Some creativity changes things overnight; some, like the innovative, continuous improvement strategies fostered by total quality management methods, more slowly. Nevertheless, reality is affected by our focused effort to improve our own lives and work. When we implement a solution, moving it into our daily routine, we affect the future.

Evaluation

Built into the implemented idea are its own evaluation criteria. Is it doing what it promised? Are profits higher and operating costs lower? Is it easy to see a higher level of productivity after the idea is in place? Are more people satisfied with the work the idea was designed to support? Does it solve a technical problem, like fixing a broken software package, extending the bandwidth of a computer communication system, adding power to data collection and analysis technology, etc.?

We can usually measure the results fairly quickly, although new ideas have varying degrees of immediate success. What the manager must do is evaluate how long after implementation is enough information available to judge whether to continue, modify, or abandon the creative solution. He must be aware that any new idea has an acceptance curve as the working population gets used to it and learns how to take advantage of it. The *New York Times* reported recently that after a decade (the 1980s) of serious attempts to streamline information management with computers, only in 1993 did any evidence of improved worker productivity emerge in the slightest measurable amount. Changing computer systems and expanding computer accessibility and power seems like a good idea, but the long learning curve and the effect on organizational culture and work habits is so severe that it might take a decade to see the positive change we are looking for. That means we must execute a combination of patience and control in observing the value of new ideas at work.

In addition, the evaluation process will likely yield new problems inherent in or actually caused by the solution itself. Do these new problems mean the solution is doomed? Not necessarily. New problems and challenges always emerge with creative solutions. These new problems are now candidates for the full, creative problem-solving process. Starting with our first awareness of a problem, we can take it, step by step, through the creativity cycle to discover and invent more ways to improve on the original solution. In this way the creativity cycle becomes interestingly complex, deeper, self-correcting, and iterative.

EPILOGUE TO THE PROCESS

Implementation and evaluation round out the creative process. Don't be quick to eliminate most favored solutions because something about them is difficult to implement or receives a negative evaluation. The real challenge is to iterate the process from the beginning, translating the implementation problem or negative evaluation into a new problem or challenge statement and taking it through the whole creative cycle.

A story is told about the manager within a major technical company who spent $2 million on a risky solution for a technical product idea that eventually failed. She was called into the president's office, presumably to be rebuked for her failure. The manager prepared her resignation letter and went to see her boss. After apologizing she laid the letter on the president's desk and asked if there could be a severance of any kind to support her while she looked for a new position. But the president replied, "You aren't going anywhere. I just spent $2 million to educate you in managing the creative process." With that, the president assigned the manager a $7 million project, which became one of the company's most successful products.

The cycle itself teaches us new ways of thinking as we become more involved in it. Another example of this is the shift in mindset necessary to go beyond devil's advocate thinking so that we are free to choose the most favored of all possible solutions. Creativity involves risk. Taking the easy way out inevitably leads backward to redo and correction, rather than onward to innovation and continuous improvement. Managers need to teach themselves, their staffs, and their colleagues to go for most favored solutions.

The irony is that individuals and organizations continually erect the most imagined obstacles in the way of the choices they most want to make. It's almost a corollary of the well-known Abilene Paradox. Presented by Jerry Harvey in a book and a film of the same name, it suggests that a major disabling factor in management is the inability to manage agreement. This is especially true when everyone agrees that a certain route is the wrong route to take, but because of imagined terrors of abandoning the route, no one dares take a new direction. Even though everyone agrees it is wrong, no one will be first to turn around or away. Similarly, we fear the possible results of the hard choices necessary to implement most favored solutions because they are often full of short-term sacrifice, hard work, and significant change. Yet we know, when forced to admit it, that it is the direction we most want to take. Creativity, therefore, requires courage, guts, and a willingness to risk and change everything for the sake of personal fulfillment and organizational success.

IMPLICATIONS FOR MANAGERS

The manager in the modern organization is a source of creativity. This is because she can foster an environment in which individuals and teams can pursue creative behavior. By providing a creative environment, managers support breakthroughs in technology and services. But how is such an environment fostered? Do you have to buy crayons for everyone's desk, or send everybody to meditation workshops? Well, some managers do, but that is a more extreme approach. The mechanisms for establishing a creative environment are at your disposal without a great deal of financial investment, since the primary ingredients are attitudinal and behavioral.

The best way to establish such an environment is to model it in one's own work style. A seasoned mentor of mine once told me that managers seldom get what they *expect* from others and hardly ever get what they *inspect* of others. What they get from others is what they *model.* Whether the manager models this by holding team problem-solving sessions using the techniques described in this chapter, or by rewarding concretely others' creative styles and outputs, or by some other unique set of motivational strategies, the manager herself is crucial to whether others will take the necessary steps to expand their own sense of themselves and their work. She is the key that says, "Be creative and grow. Don't just tow the line and survive." Managers choose not only for themselves and their staffs, but also for their whole organizations and the future of their technologies. Here are ten specific things managers can do to foster creativity in individuals and teams:

- Display your personal creativity often and openly. Be yourself and let others know you and what you do to enhance your own creativity.
- Provide real time, opportunity, and some space for individuals and teams to try creative ideas and interactions.
- Reward risk taking with praise and recognition. The manager must foster an environment in which risk taking, idea generation, and tender loving care are encouraged and rewarded constantly. She must also foster and reward innovative thinking and the gradual improvement concepts.

- Hold the whole team accountable for trying new creative tools and techniques, so that people will support each other for thinking differently.
- Remove barriers from the way of those attempting to work and think differently. For example, managers need to spend more time taking communication and bureaucratic walls down for their employees and little if any time erecting them.
- Provide fast and efficient communication infrastructures for people to use in gathering and sharing information. Use computer-based information transfer extensively, but don't abandon face-to-face conversation and interactions!
- Support creative behavior by representing unique ideas and talent up the management chain for attention and further recognition. Don't hide the lights, display them.
- Foster open thinking, not judgmental thinking, by making the rules of brainstorming the common law of communication in the department.
- Celebrate victories, however small, and involve everyone.
- Demonstrate creative behavior while leading meetings. Use group problem-solving skills and include everyone in the decision making whenever possible.

CONCLUSION

Fostering creativity and innovation is not just a nice thing to do for employees. Managers everywhere, in every line of work, and in every organization or business are called on to lead the way into new ways of working, thinking, and solving problems. Finally, the effective management of the creative process—yielding breakthrough solutions to the toughest problems facing society and its institutions—is a survival skill. We cannot go into the 21st century without it.

11 CHAPTER

MANAGING CHANGE

CHANGE IN CONTEXT

"Change is the only constant." Perhaps people say that to comfort themselves in the face of the rapid and constant changes affecting almost all organizations and individuals today. Pressures from a changing world are felt in each of our lives as never before. Over the last 150 years, our society has changed dramatically from an agricultural base in which families largely existed by their own ability to grow and sell crops, through an industrial age in which "getting a job" meant following opportunity from the family farm to the city, through the social and economic ravages of the 20th century, in which we can now see the formation of a world of work based on the flow and management of information. Technology has spawned the information age. Computers and communication devices, along with the jet plane and the personal automobile, are driving the world toward democracy and separating the "haves" from "have nots" according to who has and can manipulate what knowledge. From the stock market to medicine, information and mobility are changing the world.

In this information revolution, knowledge and the use of it are critical to organizational and individual success. Underway concurrently is a dramatic shift in the way we organize and carry out work. Complicated, hierarchical management structures in organizations are giving way to flatter, leaner, more efficient and cost effective ways of working. Driven by the evolving quality movement and by the global competition emerging during the last 50 years, organizations are constantly striving for more effective management approaches that allow each employee to be as productive as possible.

This pressure to change and adapt brings with it great stress. The stress is felt by the organization as it attempts to transform itself into a different creature in the jungle of competition and opportunity. And the stress is felt by individuals as they watch their lives, careers, and often their personal security challenged as never before.

And in this environment, the manager must manage change. Not trivial changes, but cosmic ones that redefine the organization's purposes while throwing individuals into chaos and uncertainty about their jobs, their futures, and sometimes their own identities. The situation is serious and requires great attention to detail and process if members of our work force—whether CEOs, scientists, lab technicians, secretaries, or mail clerks—are to have a fighting chance of succeeding and of helping to keep our institutions on the right track into the future.

SOURCES OF CHANGE

We live in a world of relentless and significant change. Certainly changes forced on us by the outside world are central to our experience. Often things happen at the global level that eventually bring about changes in our local environment, a ripple effect. But not all change comes from global forces or from the affect of technology on information management. The sources of change can be as varied as the situations in which we work. How to manage change depends a great deal on its source.

Most changes a manager must confront are initiated within the organization itself. Whether or not they are responses to outside pressures, they are likely to cause distress and concern among employees. Company reorganizations, lay-offs, new job assignments, shifted reporting structures, and altered demands from other departments will affect each of your employees every day on the job. Any change that affects people's finances, for better or worse, will evoke a strong reaction. Managers need to pay attention to the larger organization and be aware of changes as they surface. Just as global changes ripple out to the local level, changes in one part of the organization cause changes in other parts.

The manager of technical people must remember that, to focus on their technical work, they may screen out the politics and business of the larger organization. So a change can come as a great shock. They look to the manager to inform them, prepare them, and even protect them from the changes within the organization. A manager may not be able to do all that, but she can be aware of and help individuals respond to the changes as they occur and, with a little foresight, before they cause havoc.

Sometimes the changes that most stress employees may actually be brought on by them! As a group they may decide to implement a new policy or procedure, or change the way the office is managed, or even alter the layout of their workspace. All of these self-induced changes are matters for the manager to watch closely. Some changes initiated by the employees themselves (or by the company) actually increase productivity in the short term, whereas others may have the opposite effect. Monitoring the results of changes enacted within the organization is a key responsibility of managers attempting to keep their work forces productive.

Often change comes from within the individual. People go through personal changes as frequently as their organizations do. This is the most common arena for change management. For example, individuals experience important life changes as they age. Staff members of any given organization range from teenagers to septuagenarians. All along that range significant personal changes are occurring. The younger people may be marrying and starting families, or identifying their preferred life styles and ways of working. These changes affect morale, on-the-job expectations and relationships, and productivity. People in middle age may be experiencing personal interest shifts; what interested them during the first 20 years of their careers may become boring or even repulsive in their 40s. They may need new challenges and opportunities. In addition, with children growing up and leaving home, these employees may want more freedom and time to themselves, and may be less interested in the rat race and high pressure of work. Where the younger employee might be motivated by money and career advancement, the middle-aged employee might be rethinking priorities and looking for other ways to fit into the work culture. And of course, employees at the end of their careers may shift behavior from productive worker to valued mentor. Or they may get tired and begin dropping out.

Individuals may change as a result of life experiences and new learning. As they develop new skills and explore new opportunities, their values and expectations may change. This often leads to new life styles and work interests different from those carved out in youth. As a result of merely living, change is inevitable.

A range of changes in an individual's condition, motivation, and productivity exists within any work group, and the manager must remain aware of these shifting realities. It's almost as if the manager must understand and appreciate the changing psychologies of life to bring out the best from each member of his working group at any given time.

All of this means the manager must pay attention to each individual in the work group and develop a sensitivity to these personal changes as they occur. Is this because he is supposed to be the developmental therapist for each subordinate? Not at all. It is because he has to manage the fallout of these changes. Each individual's development affects the rest of the group, its identity, stability, and ultimately its productivity. When changes happen in an individual's life, positively or negatively, the whole group feels it.

Other sources of change exist, but these are the big ones:

1. The global conditions in which the organization works
2. The organization itself—management, other departments, and so forth
3. The working group
4. The individual employees.

CLASSIC RESPONSES TO CHANGE

The automatic response to change is to resist it. Even those changes that come from within us, we usually resist at first.

Change challenges the *status quo.* It makes us uncomfortable. Even those who boldly announce that they thrive on change cannot deny a pang of uncertainty and discomfort when the need for change first arises. Change is inevitable, of course, but think of the personal ways you have tried to delay it in your own life. When change rears its head, we feel anxious. In response to the anxiety, we turn the other way, hoping the cause will disappear. As the anxiety builds and the need for change becomes more acute, we may suffer a full-blown stress attack. Change is, after all, the major source of stress in everyone's life.

We are not managing the change when we resist it; rather, we are managing, however badly, the affect the change is having on us. *This cycle is complex and self-defeating—the more we resist change, the more acute the need for it becomes, and the more our stress level increases. Only by responding with action can we arrest this cycle.* Figure 11-1 captures the point.

Similarly, if we ignore the change too long, it becomes harder to manage any creative or successful response. Although it is important to determine the best intervention point in the change cycle, in general, *we reduce the effectiveness of our response by our continued delay.* Figure 11-2 illustrates this point.

It doesn't take much looking to see how whole organizations have resisted and delayed change as long as possible time after time. This usually leads to more pain and stress. The longer the need for change is ignored, the greater the impact of its arrival. What may have started out as a need for a small adjustment could become an entire revolution because of procrastination. By the time the appropriate responses are undertaken, the entire company may be on the verge of burn-out and demoralization.

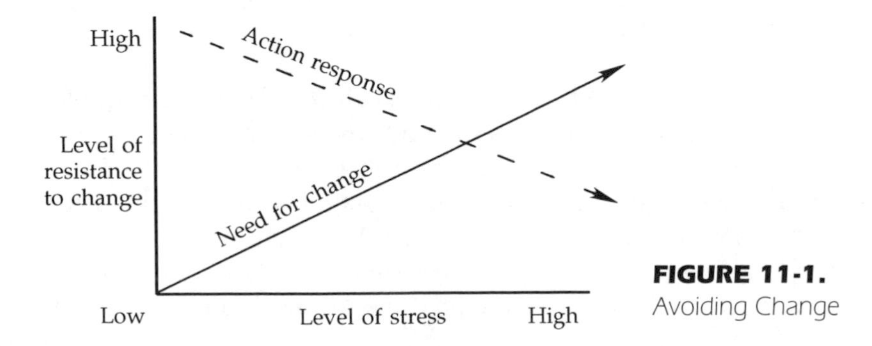

FIGURE 11-1.
Avoiding Change

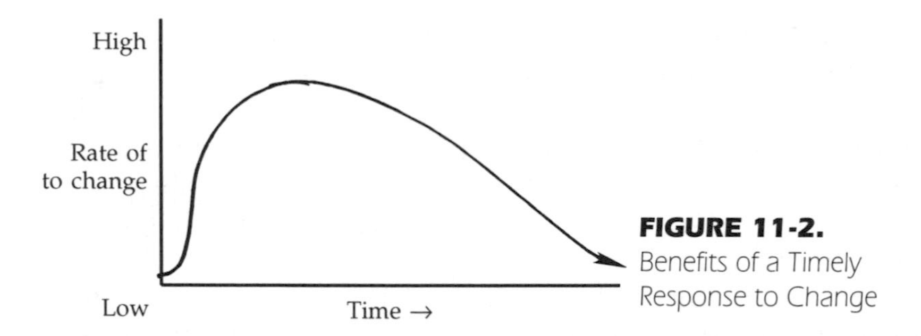

FIGURE 11-2.
Benefits of a Timely Response to Change

Since the natural response is to deny and delay, we must consciously intervene with a managed approach. Managing change has more to do with managing our attitude toward it than anything else. We can respond more creatively if we change the emotional, gut-level response from fear and threat to exhilaration and opportunity.

Of course, some change *is* threatening, and sometimes doing nothing *is* a valid response. But in the technical world where change is equated with market share, financial leadership, and technological advance, standing still is rarely an acceptable choice. In fact, the attitude adjustment needed is to trigger our flexibility and agility.

Flexible thinking and agile responses are at the heart of the creative process. Creativity is our biggest weapon in managing change. Ask questions of the individual and corporate perspective at the moment of change:

- What advantages does this change offer me? Offer us?
- How can we make it work for us?
- How can we take charge of the change and direct it for our benefit?
- What opportunities are embedded in the change that we should define and seize?

The manager can effect some of this attitudinal shift by modeling successful change management. Employees tend to be heartened and are more willing to take part in the creative approach. It is simply another opportunity for the manager to model the attitudes and behaviors she expects to see in those who work for her.

In short, we can make ourselves accept change with some willingness, and we can learn to thrive on it. We must alter our attitude toward change and begin to address it soon enough to allow us control over it. By responding sooner rather than later, we have more options and less stress to contend with.

THE MANAGER'S GUIDE TO MANAGING CHANGE

Managing change means managing our response to it—internally in our attitudes and emotions, externally in our behaviors. At times we can actually initiate change in our lives and our work. This section describes the key steps in successful change management.

☞ *Recognize early that change, or the opportunity to change, is coming.* The first step in dealing with change is to recognize it. The manager must become more sensitive to the conditions and circumstances that produce change, then help others see the coming changes to reduce the shock of transitions. By articulating change so that people can understand it, the manager provides an information safety net. The fear of what we don't know is always greater than what we can identify and understand. When we see change clearly, we can work with it. When we see it only in the shadows, it evokes the fear and avoidance we have already discussed.

In the same way, the manager needs to be more sensitive to the opportunity to initiate change. By articulating these opportunities to his work force, he can engage others in a creative, enabling process of anticipation and invention, as opposed to the demoralizing process of burn-out and mere survival. When the manager can effectively scan the horizon of opportunity and recognize possibilities for leadership, the whole work group he manages can take part and benefit. All team members can be involved in looking for change opportunities that can benefit the work effort. Every effort possible should be made to enable and educate others to anticipate and suggest possible areas of change.

☞ *Describe the change in detail.* Define the situation carefully and understand what is leading to the change. In so doing we prepare ourselves and the work group with the best possible information. The more people know, the less they "awfullize" about how bad things are getting and how disastrous the situation is. Information leads to the power to act. Secretive managers get paranoid workers. Managers who inform get workers who participate. The process of describing the change in detail, then, helps us better understand it and gives everyone maximum information with which to proceed.

☞ *Analyze the implications of the change.* Each change scenario offers different possible outcomes. In fact, one change can have many potential outcomes. Imagining the possible outcomes of a significant change accomplishes several things. First, it tends to relieve anxiety because people assume the worst. Mostly, the worst is well within the system's tolerances.

Secondly, examining the implications of the change is another way of understanding it. This offers perspective and illumination. Options present themselves even as we analyze the implications. Perhaps an impending change implies some of our work space will be eliminated. Although we will have to accept that, we can begin to see it as an opportunity, for example, to consolidate space and save rental money, begin looking for better, more efficient space, or change ways of working to streamline the need for space. Analyzing the implications of change begins a creative process. Our reactions become contributions to possible solutions instead of panic attacks.

Finally, examining implications helps to predict the outcome of change. This skill is important in the change management process, as well as in other areas related to it. For example, such predictive skills are crucial in building alternative scenarios for strategic planning. This is a business skill and a personal skill that we all need to improve as we face the ambiguous and interdependent world ahead.

☞ *Isolate and identify the areas of threat and opportunity.* We can now more thoroughly analyze the areas of threat and opportunity presented by the change. Threats might be serious, but they are better dealt with than ignored. By defining them we begin to act. As shown above, some threats can immediately be translated into opportunities. Losing work space may well be a terrible threat to our work effectiveness. Perhaps we will have to deal with that directly and fight to keep it. Identifying that reality as soon as possible gives us maximum time to build an argument or to search for acceptable alternatives.

In this process, it is better to err on the side of over-analyzing than under-analyzing. The thing to keep in mind is that it is a process, not a panic button. We consider all the threats to position ourselves to respond effectively, not to worsen our state of anxiety. Therefore, the analysis of threats and opportunities should be thorough, but not negative. In the same way, we needn't become too enthused. Don't fool ourselves into thinking that losing the work space is the best thing that could happen to us. If that's true some day, it will only be because we made it true!

☞ *Build contingency plans for threats, but focus on opportunities.* Wallowing in threats and building only defensive responses can kill creativity and leave us feeling trapped. Optimize the change by looking for areas of leverage instead of focusing on protective strategies. All successful companies position themselves and their products or services aggressively against competition by strategizing the most effective way to deal with blocks and barriers. Individuals and work groups can do this by carving out our own space in the face of change.

☞ *Employ creative thinking strategies and behaviors.* Managing change is when we most need creative responses. Whether it is to leverage opportunities or invent entirely new strategies, we need to think in new ways. Creative thinking strategies consist of gathering the team of workers together and getting everyone's ideas and input. "Brainstorming" allows us to evoke *without judgment* as many ideas and perspectives as possible. Since different members of the work group will see the changes and possible responses in different ways, the manager needs to involve everyone in the process. Ideas that interest people (whether or not they seem immediately practical) can be expanded further. Remember, in times of significant change, individuals and groups need maximum freedom to reach in new directions and try new things. Since the old rules of your working environment are being challenged, don't get stuck in your old rules of responding to it. Push the edges of perception and go for ideas and possible reactions that upset your apple cart. Be bold in the face of change. Meet the challenges that change brings with creative ideas that might not be possible in ordinary times. The bold innovation that you allow to surface now, whether a technical invention or a work procedure, may mean the difference between demise and success. So go for it.

☞ *Become part of the change, not a stubborn resistor of it.* In the martial art of Akido, masters teach their pupils to go with the energy of the attack. If an attacker thrusts his fist toward your abdomen, move into the thrust, take its energy, and move with it. Guide the thrust with your own energy to where you want it to go, in this case harmlessly off to your side. All the moves of Akido are designed to manage the energy that outside forces provide.

The same can be true of managing change. Redirect the force of the change in the ways you want to go. Use the energy of externally imposed change to carry you to your destination, and use the forces of internally initiated change to satisfy your urge to grow.

In practical terms this means co-opting the changes for your benefit by spearheading them yourself whenever possible, and improving on the changes others suggest. Building on change is better than resisting it. If I give you a good idea for a new way to do something, you might resist me directly with some classic nonstarter responses like "It'll never work," "We tried it before," and "It costs too much." Or you can say, "Yes, and let me show you a way to improve on it and better implement it." In other words, you help make it better and become part of the creative process. Remember the lesson of Akido. Even if you don't like the change, you have more chance of affecting it and changing the change itself if you are a part of it than you ever have of stopping it by force or coercion.

SUMMARY

Change is now a permanent part of the landscape. I always feel a little sorry for the person who says, "I'll be so glad when all these changes are over and things settle down and we can get back to normal." "Normal" is now change and managing change. There is no settling down.

The key is to see change as the nature of work. To embrace it and enjoy it is to let it be a catalyst for personal and organizational creativity. Out of change and creativity come new possibilities for personal and organizational growth and profit.

The final irony is that all change is evoked by us, the people who perpetrate it. We must love it as much as we resist it. And when you think about it, why not? We are all hungry for new experiences of discovery and growth, and change is more the result than the cause of our human condition. When we accept this tenet and see ourselves for who we really are, the movers and shakers of Planet Earth, then managing change is no more than managing life itself.

Do this once a year only—more often is not necessary. This is the data collection phase.

Figure 12-1 is a typical format for categorizing time data. Customize it to your needs. Once you have analyzed your time and determined where it went, ask yourself two questions: "Am I spending my time where I thought I was?" In other words, did you really know where your time went, or was it a surprise? And, "Am I spending my time where I should?" Do you spend your time on your priorities? Does your time go to those activities of greatest importance in meeting your goals?

☞ *Anticipation, not remediation.* Put another way: prevention, not correction. The good engineering manager knows that it is best to surface and solve a problem as early as possible. The earlier you surface and solve it, the less scrap and rework are involved, the less political it becomes, and the more options you have. Among the many techniques for surfacing problems in the engineering field are brainstorming, focus groups, design reviews, and simulations. Another is an old stand-by, the project review meeting.

☞ *Planning, not firefighting.* Many people say, "Who has time for planning? I've got work to do." So they're essentially reacting,

Category	M	Tu	W	Th	F	Sa	Su	Total hours %	Reasonable hours %
Customers									
Planning									
Travel									
Total Hours									

1. Necessary for routine work
2. Worked along; forward planning
3. Meetings and conferences
4. Outside meetings
5. Customer contacts
6. Special situations and assignments
7. Reading
8. Lunch (business or personal)
9. Travel, in-transit time
10. Business entertainment
11. Non-contributive (Why? You or them?)

FIGURE 12-1. Format for Categorizing Time Data

with no direction, going from fire to fire. Without planning, you cannot optimize your time. Without planning, you cannot anticipate subsequent problems. Unfortunately, in many organizations too much praise and recognition is given to firefighters, who actually were the cause of the fire in the first place because they didn't plan.

☞ *Flexibility, not rigidity.* In the final analysis, the only certainty in an engineering environment is change. Because of the technical progress we make, because of changing client goals, because of any of a number of things, our planning in general and our time management in particular must be flexible. Know your plan will have to change, know you'll have to juggle, and act accordingly. Know where you have relative certainty and can do more detailed planning. Know where you have relative *un*certainty and can do detailed planning only in the very short term.

☞ *Set self-imposed, meaningful deadlines.* Most people seem to work better under a clear-cut deadline. This is not to say that you should wait until the last minute, but you do need to know when the last minute is. It is often helpful to plan backwards from the due date to set intermediate milestones.

Know how much is too much. As Figure 12-2 implies, we can only grossly—never accurately—assess efficiency against time. As good managers we would like our group's efficiency to be as close as possible to a realistic maximum—the best that it could practically be—E_{max} on the graph. That maximum turns on one underlying question: How many tasks can a person do and remain efficient? We all know about those few people who can work on only one thing at a time. We also know of the great optimists who are jug-

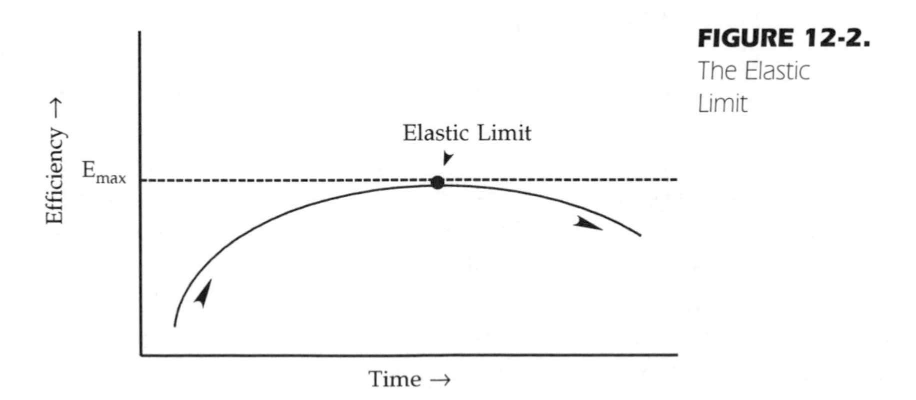

FIGURE 12-2. The Elastic Limit

gling so many balls that, invariably, half of them are on the floor. Most people, most of the time know approximately how much is too much. Experience teaches that doing more than three to four tasks simultaneously is inefficient.

The point beyond which the efficiency curve degenerates is called the *elastic limit.* This is not a management term, but one adopted from technology. The elastic limit. Technical managers must remember that groups and people are not infinitely elastic, that beyond a certain point they may actually "rupture." From the viewpoint of creativity, working on two tasks at a time is more efficient than working on one: when you hit a mental block on one task and move to another, your subconscious is still working on the first task. By alternating between two tasks, you can be extremely creative. Often three tasks are possible at the same or better efficiency; with some people and some tasks, four can work, too. But from the standpoint of efficiency, going above three or four can cause a rupture.

☞ *Batch similar tasks.* Working on tasks with some commonality offers an efficiency of scale that increases your elastic limit. The commonality may be similar technology, the same client, the same general specifications, or the same manufacturing process, to suggest a few examples. So you can move from one task to another with minimal mental dislocation or wasted physical motion. Many of the changes made by organizations in the 1990s are based on the commonality or batching approach. For example, a group whose projects are all from one client will save time and money from fewer different specifications, fewer customer liaison points, fewer new relationships to be established and nurtured, greater consistency of policies and practices, and so forth.

☞ *Effectiveness.* Work on the right task at the right time. Efficiency is input compared to output, or doing things right the first time. Effectiveness is timing tasks and doing the right things first. To manage your time well and to plan your projects well, it's important to understand the time phasing and dependency (interrelationship) of various tasks. The best way to understand this is to draw a network diagram of your tasks. The interrelationships required by the flow of work on the project become clear. Network diagramming is strongly recommended for all, and not only for scheduling purposes. Preparing a network diagram in rough format will help you understand the tasks you have to do and the interrelationships among them.

☞ *Prioritize!* This is important. You don't have enough time to do everything. You've got to decide which tasks, if left undone for the moment, will not really hurt you. This is called benign neglect. This can be a tough call, but it's critical to making sure important tasks are accomplished and avoiding major lapses.

☞ *Delegate!* You have two options: Delegate to the lowest level with *absolute* certainty that the task will be done perfectly. Or, delegate to the lowest level with *relative* certainty that the task will be done perfectly. After thinking about it, most people will want both. Perhaps the 80-20 law (Pareto's Law) comes into effect here: delegate 80 percent of the time to the relative certainty level and 20 percent of the time to the absolute certainty level. Go for absolute certainty on critical path items, irreversible items, extremely expensive items; but for the vast majority of items, accept relative certainty. If you require absolute certainty in all tasks, nobody will ever grow because they will always be working at their level of certainty. Only with some degree of uncertainty and risk do we truly grow.

☞ *Visibility.* You can't do what you can't remember. Too many people prepare a schedule at the beginning of a project, then let it sit in their desk drawer. Once a month at the project review meeting, they pull it out, dust it off, and use it to play the game of "So, what happened this month?" If you want schedules, budgets, and specifications to have meaning, make sure they're visible to everyone all the time. There is no alternative. One of the authors of this book insists that schedules be posted on all four walls of each office. No matter where you look, you're faced with the fact that you have a schedule to meet. There's no escaping it.

☞ *Consolidate communication.* Don't visit the same people ten times a day. Try and consolidate your business with these people. You'll save transit time and preparation time, and planning what you want to discuss with these people will give you another opportunity to over-view the project. For most people, seeing them once or twice a day is all you'll ever need. Don't impulsively get up from your chair and dash out of the office every time you think of something to say to your associate.

☞ *Control interruptions.* This may be the most important point of time management. An hour of uninterrupted time is worth *X* hours of interrupted time—from 2 to 8, depending on who you ask. If you can find an hour of uninterrupted time during the course of a day, you will gain at least two extra hours. That is why controlling interruptions is so important. Following are some of the methods people use to control interruptions.

- *Quiet hour.* Select an hour during the day when you schedule no meetings and accept no calls or visitors. For that one hour, the group is *en vacua*—in quarantine, closed to the public. Obviously, this method is limited in application. If you have a heavy interface or liaison role, or production calls to tell you the line is down, you can't say, "Sorry, it's quiet hour. We can't speak to you now." Or if a customer calls with a question or wants to place an order, you can't say, "Not now, it's quiet hour." But for many groups, it is a good technique.
- *Close the door.* Many of us don't have a door or are in a multiperson office. If you do have a door and can close it, this may indicate to others, "Please don't interrupt me." If you find that, in your office, a closed door invites someone to bang on it and yell, "Are you in there?", try hanging a graphic on your door that means "Don't even try. Death to the intruder." Maybe a skull and crossbones, a high voltage sign, or a hazardous material sign—just keep it light. You want to convey, "I'm very busy. Make believe I'm not here."
- *Hide.* This may be the most popular. A lot depends on how large your operation is. It's a lot easier to hide in a million square feet than in five thousand. Find an empty office, maybe that of a friend who is on vacation or traveling, that is far, far away from your area. Empty conference rooms are useful, particularly if you have a lot of drawings or other material to lay out. The company library, if you're lucky enough to have one, is a good place. (Unfortunately, it's often the first place people look for you.) Special test areas not in use (such as shock and vibration areas, temperature and altitude areas) or cafeterias or canteen areas in between coffee breaks and meals also work. Use your ingenuity. Remember, don't use the same place all the time, or else people will know where to find you.
- *Stay home.* Many people are shocked to hear that some find staying home the most useful method of controlling interruptions. Obviously, it won't work in every situation. You

can't use it in military security work or with proprietary information. You can't use it when you need access to heavy machinery or other items that must remain at work. But for a lot of tasks, staying home might be your best bet. People who use this method do so as a last resort. A report is due tomorrow. You've tried everything and you can't get it done. You tell your boss, "This is the only way I can do it," and you stay home.

Staying home is not for everyone. You have to be very, very self-disciplined, and you have to have no distractions at home—either other family members gone or very well trained! Given these conditions, you can concentrate and get your work done without interruption. The phone is available if you need to get some information. If you have a computer or a terminal at home, maybe you can hook into your office system. Those who have all this find staying home an excellent trick.

- *Offset your lunch hour.* If, for example, the common lunch hour in your organization is 12 to 1, and you need an hour to prepare for a meeting or to write a memo, try doing that from 12 to 1 when everyone else is out, then have your lunch from 1 to 2. Going out from 11 to 12 may not work because everyone will know what you've done.
- *Come in early or stay late.* Some might say come early *and* stay late. All of us have a personal energy pattern. About two thirds of Americans reportedly have high energy in the morning and are bottomed out by 5:00. The other third are just the opposite. They are sort of catatonic first thing in the morning, but by 5:00 they're up to speed. Use this to help you manage your time.

  If you have high energy in the morning, try to do your most difficult tasks then. If you're a night person, do them at night. If you have to work a few hours extra today and you're a day person, don't stay late tonight when you're working with the least energy; come in early tomorrow morning instead. Incidentally, most of us (both night and day people) have a low energy dip during the course of the day; we usually notice it right after lunch, but some research suggests it's really 12 hours after the midpoint of your night's sleep (i.e., if you sleep from 11:00 to 6:00, you would bottom out around 3:30). Some people say that the best time to hold a meeting is right before lunch. The stomach becomes a powerful accelerator to finish the meeting.

Now that you have a feel for key concepts of time management, study Tables 12-1 and 12-2 for ideas you can use to improve your management of time.

ABOUT MEETINGS AND TIME

☞ *Be punctual to meetings.* In many environments lateness is a way of life. If this happens where you work, take work with you that can easily be put aside.

☞ *Be certain that those attending know what the meeting is about by*

TABLE 12-1.
20 Ways to Make Better Use of Your Time

1. Reduce your paperwork.
2. Delegate.
3. Select one portion of the day as "not to be disturbed."
4. Do most of your telephoning at one time.
5. Learn to say "no."
6. Plan the use of your future waiting time.
7. Ignore as many trivial requests as possible.
8. Have a policy of not doing things.
9. Avoid meetings where possible.
10. Be a clock watcher; keep a time log.
11. Let George (or Georgia) do it.
12. Throw out everything possible every day.
13. Know the purpose or function that you are trying to achieve.
14. Know and use Pareto's Law.
15. Do the right thing before doing things right.
16. Schedule and allocate work in the most effective manner.
17. Perform efficiently those activities that you must do.
18. Stop doing things that you can do best.
19. Mechanize routine things where possible.
20. Don't turn decisions into research papers.

TABLE 12-2.
20 Time-Saving Tips

1. Arrive early or stay late.
2. Write a daily "to do" list, or keep a perpetual one.
3. On Monday morning (or Friday afternoon), plan your week's activities.
4. Have your secretary screen your mail and/or phone calls.
5. Jot down spontaneous notes and ideas; don't lose them.
6. Cut off nonproductive activities (e.g., long calls).
7. Write letters to people who talk too long on the phone.
8. If you dictate to machine, try a stenographer, and *vice versa*.
9. Set aside your most productive time to do creative work. Use your low-energy period for trivial, routine tasks.
10. Speed-read where possible.
11. Keep schedules visible.
12. Handle every piece of paper only once!
13. Answer some letters by writing on them.
14. Have a place for everything; avoid searching.
15. Listen carefully; take good notes.
16. Try doing your thinking on paper.
17. Is your boss a reader or a listener? Communicate *via* his or her preferred way.
18. Set deadlines for yourself.
19. Carry a bound pad or notebook with you.
20. Try using a mini-recorder as an electronic notebook.

preparing and distributing an agenda. Show who is invited, what material should be updated and brought to the meeting, who is expected to address each agenda item, and how long the meeting, and perhaps each agenda item, should last. Preparing an agenda has the added advantage of ensuring that *you* think carefully, one more time, about the items on it! If you are invited to attend a meeting and there is no agenda, it may behoove you to call the convener and get this information beforehand.

☞ *Think carefully about whom you invite to a meeting.* Try to save time by inviting only those who can contribute, those for whom it is part of training, or those who have a need to know. If in doubt appraisal technique on the video is better than what they have experienced.

Six types of performance appraisal are in common use.

☞ The *matrix format* resembles a grid with gradations of performance—good, average, bad—on the horizontal scale and performance factors—meets budgets, meets schedules, new business development, and so forth—on the vertical scale.

☞ The *forced curve* is similar, except that a forced curve is used for the gradations; that is, in a five-part system—outstanding, good, average, not so good, terrible—the curve would force 10 percent into the first category, 20 into the second, 40 into the third, 20 into the fourth, and 10 into the fifth. Incidentally, we have found that the first and the last categories are rarely used. Research in questionnaire design and in performance appraisals has found this, too. Therefore, if you really want a system that uses five categories, you should design *seven.* Go figure.

☞ *Management by objectives* first became popular in the 1960s. Based on a new philosophy of management, supervisors and subordinates mutually agreed upon a set of goals that the subordinate would try to meet during the coming review period. The goals were selected to reflect the personal career goals of the employee, the group goals of the supervisor, and the overall business goals of the organization. The employee's next review would be based on her or his progress on these goals, however they may have been revised in the interim. MBO's popularity waned in the 1970s because it became an administrative nightmare: some organizations had 25 goals with mid-course corrections every month, and in many engineering environments tasks change all the time. A simplified MBO system reappeared in the late 1980s, and unless your area's goals are continuously changing, this simpler system is good for the engineering environment.

☞ *Forced ranking* operates this way (brace yourself): As a first level manager, you rank your six subordinates from 1 to 6, "best" to "worst." Make these evaluations however you may, perhaps by hearkening back to the first system. Then you and the three other first level managers sit down with your boss, and you merge, or rather your boss merges, the four different groups into one forced ranking for the section. This usually ends up in a bloodbath, all the first level managers yelling and screaming for their own people.

Next, your boss, the section head, takes the merged ranking of the four subgroups into a meeting with other section heads. Another bloodbath. And so on. Forced ranking has great value at the lowest levels where the people intimately know their subordinates, but it loses significance at higher levels: apples and pears and oranges are being evaluated by people who really don't know the produce. This widely used system is particularly attractive to personnel people because, from their standpoint, it is simple and they think it would stand as a legal justification for reductions in force.

☞ On the contrary, the fifth system is *pure narrative,* pure essay, nothing comparative, nothing quantitative. It is usually the one that the personnel people like the least because no statistics can be garnered from it, and they feel it could be difficult to justify certain administrative actions taken based on such a system.

☞ The *critical incident* method looks almost like an indented specification format in which you list major topics and, within each one, list specific critical incidents that illustrate how the person has performed.

Each system has advantages and disadvantages. Although many engineering managers are frustrated by their company's system, you should try to optimize whatever approach you have to use. The most important phase of performance appraisal in its formal sense is really the one-to-one interview between you, the manager, and the person being evaluated. And don't forget that the good manager remembers the informal system that occurs all during the year. Nothing should come as a surprise to the appraised employee during a formal appraisal because he or she should have heard about it during the year. The formal appraisal should be a distillation or a summary of the various informal appraisals (see Figure 13-1).

From the date of a performance appraisal, we should look back in time to evaluate performance during the past appraisal period.

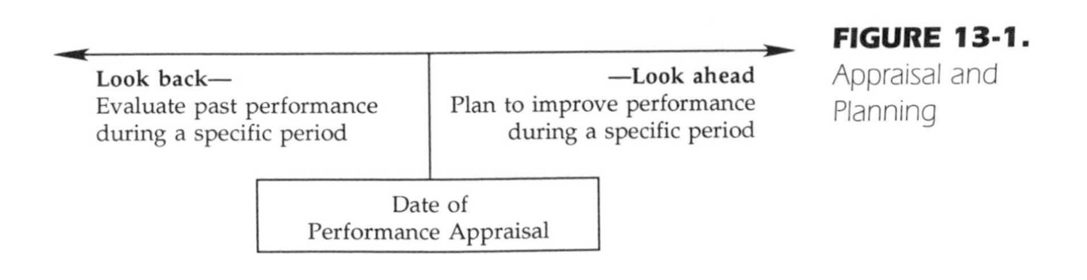

FIGURE 13-1. Appraisal and Planning

the employee should be doing, and on the right, how the employee sees the job, things the employee believes should be done. The shaded overlapping area is the area of common agreement. One goal of performance appraisal is apparent in the bottom: to increase the area of common agreement.

Engineers are familiar with instrument calibration programs: the tag on the instrument tells you not to use it after a certain date, but to return it to the instrument lab for recalibration. So it is with performance appraisals. During performance appraisals, we recalibrate our standards of performance for the upcoming period. You don't tell a person at the end of the appraisal period what the standards are. They're not supposed to be a surprise. Rather, you appraise past performance based upon standards established at the beginning of the appraisal *period just ending.* As you go ahead with performance planning for the *period just beginning,* review past standards and make any necessary changes so your employee knows what you expect and what goals will satisfy you and the organization. Your employee's performance against these revised standards will be appraised during the next performance appraisal. No surprises! If your business, your employee's job assignment, or other factors change, immediately adjust the standards with your employee.

Improving Performance

One of the most important parts of performance appraisal is determining what will improve performance. ☞ Of the three major factors, the person's *ability* comes first. We have to try to improve the person's ability; Chapter 14, on training and development, provides a review of various methods. An important component of any performance appraisal relates to career development for the employee, not only in terms of her own goals, but also in terms of her relationship to the group and to the organization. Consider this in both the long and short term: improving a person's ability makes it possible for her to perform better in the future.

☞ Improving ability depends upon the second of the major factors, *opportunity.* The person has to have the opportunity to use her ability. So unless your employee has enough tasks that make use of this knowledge, she may become rusty or out of date in these areas. One question that you must consider when assigning work is, "Are we keeping this employee up to date in the areas in which I need her as one of our resources?" Whenever you have a period of time in which you don't have work in a certain area, but that area is part of your continuing, long-term interest, consider scheduling meetings, discussions, seminars to maintain your knowledge.

☞ Finally, performance improvement will depend upon the *motivation* of the individual. As was demonstrated in Chapter 4, a lot depends on you, the manager, and on the ambience and the environment that you create. So if you wish to improve performance, you've got to improve a person's ability, you've got to give her the opportunity to use that ability, and you've got to motivate her sufficiently so she maximizes that opportunity.

PERFORMANCE APPRAISAL FACTORS

In many performance appraisal systems, the performance factors are preprinted on the form. They may include things such as quality, timeliness, initiative, adaptability, communication, and so forth. It's important to remember that you will do the best job on performance appraisal if you can use the factors appropriate for this person's job and job assignments during the appraisal period. For example, if a person is working on a one-person job off in a corner, is communication important? Is cooperation important? Not as much as creativity and timeliness. But if the person is next assigned to manage a project of fourteen engineers with two subcontractors and difficult clients, cooperation and communication become extremely important. So use the relevant performance factors. Learn to weight them; they are not of equal importance. On a form with five preprinted factors, it appears that each is worth approximately 20 percent of the total. But because of circumstances like those noted above, those five factors will not contribute equally to the total. Develop a weighting profile for the employee, the job assignment, and the mutually agreed upon expectations.

SELF-APPRAISAL

About one third of American organizations use self-appraisal formally. When it was first suggested a number of years ago, most managers thought that self-appraisal would be a major fiasco, in that individuals would give themselves unrealistically high ratings. On the contrary. In practice, about 80 percent of the people appraise themselves roughly the same as their supervisors, if not a little more harshly. Only about 20 percent appraise themselves too highly.

Two variations of self-appraisal are used. In the first, both subordinate and supervisor prepare appraisals, usually using the same format. In the appraisal meeting, they see each other's comments for the first time and compare to determine where their opinions differ. In the second variation, the subordinate prepares the self-appraisal and gives it to the supervisor some time before the appraisal meeting so that it becomes one of several inputs the supervisor uses in preparing the review. The advantage to the second method is that the subordinate has the chance to emphasize his best accomplishments prior to the superior's preparation. His input may be more influential in this variation than in the first, in which the employee may find himself convincing his supervisor that the supervisor's initial preparation was erroneous.

In those two thirds of American organizations that don't formally incorporate self-appraisal in their performance review systems, many managers use it informally. When asked, they say it is a valuable adjunct to the process. Its popularity rests on the fact that it is one of the few facets in the performance appraisal system that is a win-win situation. For many subordinates, it is the first time they really have looked themselves in the mirror and tried to evaluate themselves—a valuable experience. And having done so, the subordinate is far more equipped to have a meaningful dialogue during the performance appraisal interview. Without having conducted a self-appraisal, the subordinate walks in cold and sits down, and what you have is a monologue with an audience of one.

We recommend that companies formally incorporate self-appraisal and that managers informally use it as a way to improve the entire process. If a supervisor's and subordinate's appraisals differ greatly, it may indicate that the supervisor didn't provide enough informal feedback during the review period, or never provided clear expectations to define satisfactory performance. And if the subordinate's self-appraisal is much lower than the supervisor's, it may indicate that your subordinate has a poor self-esteem problem.

APPRAISING THE MANAGER

Another innovation that a number of managers use is having the subordinates appraise their managers. Relatively few companies do this formally, but quite a number of managers want feedback from their subordinates on how they're doing. Some do it simply by discussion. Others have devised a written format; still others do it in a group session with the staff. We encourage managers to look into this, with this consideration: although we think self-appraisal is a win-win situation for all, we don't necessarily believe that getting the employees to appraise the supervisor *when not required by the organization* is a win-win situation. If you can't be sure that someone's comments wouldn't stick in your craw, then it may not be for you. But if you have a good relationship with your staff, can remain objective, and will not be upset by the results, we think this is something from which you can greatly benefit, whether or not such a thing is required by your organization.

LOCAL PERFORMANCE APPRAISAL CULTURE AND OBJECTIVITY

The local environment and your corporate culture has a tremendous impact on the performance appraisal process. How performance appraisal operates in your organization may differ greatly from what your review form implies and what your procedure manual explains. For example, if you try to be intellectually honest and don't take into account the fact that corporate culture records "average" as "above average," you may be doing a disservice to your staff. Unfortunately, in the majority of American organizations, we discover that games are played with the performance appraisal process. It's important for managers to know how curves are developed and used and what terms like "above average" and "average" really mean. Can you say something constructively critical about one of your people, yet not have it used by your management as a reason against the increase you want for that person?

REAL-TIME NOTES

When preparing to conduct a performance appraisal, most people sit down with a blank form and start to think about all the things the person did during the course of the year. Well, naturally, the disasters surface first in your mind. Then, what she's currently working on. But what she might have been doing eleven months ago, particularly if there are many small projects, presents a problem. To save you preparation time and to make it a fair and equitable appraisal, we suggest the following: For everyone in your group that you have to appraise, prepare a file folder. During the course of the year, as both positive and negative things happen, summarize the incident or event on a file card, date it, and drop it in the folder. Then, when preparing for the appraisal, put these file cards in chronological order, and you will have an overview of the entire year based upon real-time inputs. The more often you add to the folder, the more complete your overview. If you require each of your subordinates to give you periodic written progress reviews, keep copies in the file folder as another source of real-time information.

CHAPTER 14

TRAINING AND DEVELOPMENT

We can divide the training and development methods into two broad categories: those that are part of the work assignment and those that are not. All, of course, are associated with an employee's relationship to the company or organization. They are shown in Table 14-1.

TRAINING AS PART OF THE WORK ASSIGNMENT

Special Assignments. An employee receives a special assignment not because he is the best person for the job, but because the person and the company will benefit by his gaining this experience. The manager has decided to take a short-term loss in efficiency in exchange for a long-term gain in skill base. Special assignments are given as part of a career development plan for an employee, when work of a certain type is expected to increase beyond the staff's current capacity to do it, under a hiring freeze when a unit must develop its in-house capabilities, or as part of a discussion with the individual during a performance appraisal nization have a good in-plant training program. Don't overlook lunch-time sessions. If you don't have time for a two-day seminar, break it up into several late afternoon or early morning sessions, but make sure there's a continuing training program on both technical and nontechnical issues.

University Courses. University courses are available in most urban and suburban locations. As a manager, you should keep a record of what courses your staff takes—the schools, topics, and instructors—so you will know which ones to recommend in the future. Stay abreast of what's available. Get on a mailing list for course catalogues, continuing education catalogues, and announcements of special events that might be of interest to your staff. Provide a bulletin board and/or reading rack just for this type of current information.

Professional Associations. Professional societies are a major source of training, particularly in the United States. Arrange for corporate memberships in the professional societies appropriate to your work. The national meetings, usually once or twice a year, include not only technical and management sessions, but also a day or two of tutorials on current topics. Local chapter meetings often present tutorials or sponsor special events such as interactive participation in conferences *via* satellite transmission. Through their publication departments, many professional societies provide video programs, books, and other training material. Make sure that you know what's available from the professional societies represented in your organization. Put their announcements where staff have easy access to them.

Trade Associations. These organizations consist primarily of member companies in a given industry. They often offer training programs focused on that industry. They are similar to the professional associations, but far more tunnel-visioned. Their national conventions will offer lots of training, and their publication departments will have lots of material of value to you.

Programmed Instruction. This refers to computer-aided instruction in which you learn through tutorials.

Audio-Video Programs. Your local college or university engineering library will have access to AV instruction and to associations that develop, collect, and disseminate AV instruction programs. The Association for Continuing Media Education for Engineers maintains a master catalog of the video program output of the various U.S. schools that provide this form of training. The video programs range in depth and complexity from mere briefings to those appropriate for college credit. They are particularly good if course availability is limited in your area or for people who cannot get away when courses are offered.

Management Education. A wide variety of programs are offered by various organizations in the field of management education. Sources range from the extension departments of state universities to professional management organizations to individual management consulting firms. Training can include seminars, a prepared take-away kit for you to conduct in your own organization, training the trainer, or video programs combined with books combined with other instructional materials.

Group Programs. Educational programs such as simulation games are another form of training. These include various games (both on and off computer) that help you understand certain elements specific to your organization, such as a revised financial report format, project management software, and so forth.

Planned Reading. The importance of a good company library, and perhaps a section library, cannot be overstated. Pertinent journals, trade publications, and books are essential. In addition, it's necessary to have access to the various computerized database services with bibliographical abstracts and report information. If there's a college or university library in your area, facilitate access for your staff.

- One team or crew that changes or disbands at the end
- One configuration or research path, different with each project.

Uniqueness may come from the fact that the next project will have a different usufructuary (or client). Examples of typical kinds of projects are shown in Figure 15-1.

Projects Are Temporary

Projects have a specific beginning and end, even if the people who were there at the beginning don't get to see the end, and projects begin and end at ascertainable times. Teams change as projects go through their phases, but project and team do come to an end.

This contrasts with the functional organization, which is there as long as the entity exists. Even a closed factory may still need parts of an accounting function or maintenance long after everyone else has gone.

New product research, development, and introduction
Electric vehicle
Pharmaceutical, drug
Consumer product (perfume, pasta)
Line extension (perfumed soap, low-calorie pasta)

Establishing and introducing new methods
Mandated compliance program
Anti-harassment policy
CAM/CAE/CIM introduction
Strategic plan

Construction
New building
Renovation
Plant layout change
New machine

Environmental compliance
Site clean-up
Hazardous substance control
Transportation policy and method
Regulated substance handling

Software and systems
Project control system
Programmer documentation policy and procedure
Scheduling system
Standardized nomenclature, numbering system

FIGURE 15-1. Examples of Projects

Projects Create Conflict

Projects create conflict by their nature. One conflict is between the project organization and the functional organization. Because projects are temporary, so is their organization. Yet each project calls upon the functional, permanent organization to supply or forego resources on behalf of the project. That causes conflict.

Projects tend to command attention. Project people may be signaled out for commendation, while the permanent complement—the functional people—have to do the pedestrian stuff, equally important and equally difficult. Jealousy erupts between project people and functional people. Functional managers describe project people as *prima donnas,* and project people decry the "old fogies" in functional departments.

Projects also create conflict because each project represents change in the organization.

Projects Evolve through Phases

Projects may have been hatched in a bar, at the company picnic, or in the boardroom, but all must start with a definition phase and end with the application, fabrication, or implementation phase. Most projects have four phases—concept, definition, realization, operation (see Figure 15-2).*

Projects Are for the Usufructuary (or Client)

User and usufructuary are not functionally the same. "User" is who will *use* the product or process developed for or at the request of the usufructuary or client.

The usufructuary is the person or group (an individual, an entire company, or the whole government) either that commissioned the project or on whose behalf it is undertaken. Sometimes the user is the usufructuary's employee, as when the Manufacturing VP asks that a new machine shop be built. The usufructuary may be an outsider who has ordered your product or service. In

---

*Some entities, such as the U.S. defense establishment, have five-phase projects: "threat" is added as the first phase. Commercial projects handle this through goal definition.

CONCEPT	What we are doing. Why we are doing it. Project macro plan, specifications. Standards, conditions of acceptance. Regulatory conditions, limitations, approvals.
DEFINITION	Project breakdown, structure, activities, tasks. Detailed specifications, quality assurance. Action plans, budgets, schedules. Task and budget assignments. Reports, controls. Regulatory reporting.
REALIZATION	Functional descriptions, test specifications. Work orders, prototype and breadboard models. Systems investigation and design. Internal/company site testing (alpha), tryouts. Pre- and clinical trials. Debugging, modifications, re-specification. Design and process reviews. Hand-off to manufacturing, processing. Regulatory compliance.
OPERATION	Final configuration, process, GMP. Change procedure. External/customer site testing (beta), tryouts. Clinical trial and reports. Manuals, training, hand-off to usufructuary. Repetitive manufacturing help.

FIGURE 15-2. Project Phases

many cases, the word "client" can be exchanged for "usufructuary."

You work for the usufructuary (or client). There may, however, be layers of usufructuaries.

Layer Cake. "I talked to Dr. Limpio," a sales representative with BCD Company tells the chief of R&D. "You know, he's the guy who re-attaches toes." "Yes," the R&D chief responds, "I heard about him. What I wouldn't give for him to ask us to make one of their microsurgical products." "As a matter of fact," counters the salesperson, "he has. Dr. Limpio wants a self-cleaning instrument tray and wants to know, can we make him one?"

In due time, BCD Company decides to try to produce a line of such trays. The marketing department, specifically a product manager, is given the job of driving the tray project and becomes the primary "usufructuary" to other project team members. Their work must satisfy her.

Multiple Usufructuaries. Dr. Limpio will be asked to provide specifications and conditions and, from time to time, will be asked to review progress on "his" tray, which might even bear his name coupled with that of our entity: the BCD-Limpio Antigerm Tray. He will thus be the secondary usufructuary: BCD Company's work must satisfy his needs.

To the extent that the tray will be subject at least to scrutiny by a government regulatory agency (in the United States, it is Food and Drug Administration), the agency or its representative will be a tertiary, but crucial usufructuary in that the collaboration between Dr. Limpio and BCD Company must satisfy their regulations.

And of course, if nobody buys the tray, all the time and money will have been wasted. So the product that results from Dr. Limpio's specifications, BCD Company's work, and the relevant regulations must satisfy the surgeons, the ultimate usufructuary.

Know Thy Usufructuary. The project manager must be familiar with the client because the project's standards of performance and acceptance will be either set or agreed to by the client. Those standards shape the project. Do *not* begin work on a project without clarifying the client's standards of acceptance.

Not to mention that the client or its delegate—even an uninvited delegate such as the building inspector—will appear, if at no other time, on the project's last day. On that day it would be nice to hear the magic words, "Nice. Just what I wanted."

Projects as Boats

A project can have at least one of the characteristics of a pleasure boat: "a hole surrounded by water, into which you pour money." Projects can also resemble an open vein through which lifeblood flows out and infections come in.

For all these reasons, organizations that engage in projects start their supervision over a project when it is only a glimmer in someone's eye. They ask that projects be set up and submitted for review and approval.

ESTABLISHING A PROJECT

Projects must be set up; they must be posted on the books of the organization. If the project is for an outside client, it can be set up through a Request for Quotation, Request for Proposal, or similar mechanism. If it's for your R&D chief who wants a new clean room, it can be set up through your capital budget or other appropriation mechanism.

Depending on the client, project size, risk, and other conditions, most projects are set up in two stages:

- *Preliminary*—A feasibility study to make sure that the organization should engage in the project
- *Approval*—Approving the feasibility study, the details of the project, and the project itself: costs, resources, objectives, benefits, profitability, schedule, and so on.

Preliminary Stage

The preliminary stage is a fishing trip. "There is, you see, this idea. . . . Why don't we take a look at it?" Or, "Cromfasdo Corp is asking us to bid on one of their jobs. I wonder. . . . " In this stage, possible projects are measured for size, fit, and so forth. These two streams of objectives are reviewed:

- *Purpose*—What are we doing?
- *Goal*—Why are we doing it?

Figure 15-3 shows a form that can be used to document the preliminary stage. The form has two major parts, one for general information about the proposed project, the other for information about the feasibility study. This is about as simple as it can get.

For some projects, however, this is only a summary page. Where risk is high, such as in the case of the microsurgical tray, even the preliminary stage calls for more paper, especially proving how you know what you know.

Approval Stage

After the feasibility study gives the project a positive indication, or shows that responding to the Request for Quotations is worthwhile, the real work begins. The project must be broken down into sufficient detail to prove out the budget and the schedule, support

PRELIMINARY PROJECT REQUEST	Project name

Requested by (name, title, location, phone)	Date

Describe objectives—What? Why?

Check appropriate goal summarization:

- ☐ Profit enhancing % ________ $ ________
- ☐ Increased market share % ________ $ ________
- ☐ Cost reduction % ________ $ ________
- ☐ Retire existing product. Current revenue % ________ $ ________
- ☐ Add new product. Addition to revenue % ________ $ ________
- ☐ Other ________
- ☐ Profit maintaining
- ☐ Facilities maintenance
- ☐ Mandated by law
- ☐ Safety, security
- ☐ Retirement of facility
- ☐ Support automation
- ☐ Remedy deficiency

Describe the goal in detail. Attach calculations, rationale, etc.

Describe the feasibility study of this preliminary request

Who will conduct the feasibility study (name, title, location, phone)?

Describe the deliverable

COSTS Cost item	Labor	Material	Other	TOTAL
TOTALS				

Request prepared by	Request approved by

Start date	Finish date	Why are these dates significant to the project?

FIGURE 15-3. Preliminary Project Request

the profit calculation or benefit forecast, and otherwise demonstrate that the project has value, considering the risk.

Risk. Four kinds of risks exist in projects:

- *Conceptual*—Can we [even] think of a solution to the problem?
- *Operational*—Can we produce the item? Can we make it, reasonably?
- *Marketing*—Can we sell it? Will the voters go for it?
- *Financial*—Can we profit by it? Will society benefit from it?

Risk should bring benefit. In the case of a drug, the risk may be that the government never approves its use or that the price will be too high to allow profit. Thus years and millions of dollars in development down the drain.

On the upside, the drug may be the cure of the common cold. The company will be swimming in revenue and not only reward the investors for their many years' wait, but also pay for research into other elusive cures.

The electric car risks that the battery will be too heavy, the driving radius too small, and the battery charging industry won't be ready when the car is. On the other hand, the first company that comes out with a viable electric car ... well, just think of it!

Objectives. As early as the preliminary stage, management must be told what are the objectives of the project. These objectives will be familiar:

- *Purpose*—What are we doing?
- *Goal*—Why are we doing it?

"Mrs. Kratchitt, why are you fixing eggs for your husband?"
"They're full of cholesterol. It'll kill him."

"Mrs. Kratchitt, why are you fixing eggs for your husband?
"He loves eggs on weekends and I want him to be happy."

Same eggs, different goal.

"Say," the president tells the management information chief, "have your people

"Say," the president tells the management information chief, "have your people

write a program to time and register who comes in and goes out of our place."
"Why?"
"I'm gonna set up a blacklist. Come raise time. . . . .

write a program to time and register who comes in and goes out of our place."
"Why?"
"Our access controls have been criticized. The government may close us down if we don't get a handle on comings and goings."

The difference goals make.

There are three goal-related rules:

1. Ascertain (or, if it is your project, state) the goal.
2. Document the goal.
3. Involve others in your goal.

Goal documentation is part of the preliminary stage, as we saw in Figure 15-3, and is definitely part of the approval stage, as shown in Figure 15-4. As with all forms shown here, this is but one example.

Handling the Form. Forms such as these are recapitulation, summary, or "face" sheets. They summarize what may be a pile of data beneath them. The forms pretty much speak for themselves.

This format shows space for two sets of signatures. One is from key people on the requesting side:

- *Proponent or Requestor*—We are the ones who want it.
- *Engineering or Research*—We can conceive it.
- *Manufacturing or Operations*—We can make it.
- *Sales*—We can sell it.
- *Finance*—We can demonstrate its profitability.

These signatures are affixed as the form moves up the ladder. On its way down, it gets the second set of signatures, from corporate equivalents:

- *Group officer, President, Board*—We give informed consent.
- *Administration*—You are using the correct procedures.
- *Finance*—The funds are available and authorized.
- *Purchasing*—We'll find suppliers.

CONSENT TO EXPENDITURE		Project number
Subsidiary		Requisition number
Unit, department	Location	Plant, building number
Title		Amount $
Summary, description, justification		

AUTHORIZATION SUMMARY		INVESTMENT AND CASH SUMMARY	
Material, supplies	$	Gross investment	$
Labor			
		Cash requirement	$
Operating expense			
Contractor			
TOTAL	$	Contingency	%

This authorization ☐ is ☐ is not included in the capital budget.

	CLASSIFICATION: Check appropriate box(es)
Economic life of project (years)	☐ Profit adding
Estimated start date	☐ Cost reduction
Estimated finish date	☐ Expansion or addition ☐ New product ☐ Existing product
Percent yield on gross investment	☐ Profit maintaining ☐ Quality improvement ☐ Replacement
Payback time (years)	☐ Health, welfare, hazard ☐ Government regulations ☐ Lease

Signature	Date	Signature	Date
Prepared by		CONSENT	
APPROVED BY SUBSIDIARY		Marketing	
Sales		Operations	
Manufacturing		Purchasing	
Engineering		VP Finance	
Accounting		VP Administration	
Division, unit manager		Group officer	
Group VP		President	
Board of Directors		Chair of the Board	
		Board of Directors	

BASIS OF FINANCE (check one):
- ☐ Internal generation of funds
- ☐ Increase capitalization
- ☐ Foreign borrowing with guarantee
- ☐ Foreign borrowing
- ☐ Loan from parent

FIGURE 15-4. Consent to Expenditure

- *Operations*—We agree that it's doable in the way you propose.
- *Marketing*—Yes, the market exists.

Signatures do not merely indicate approval, but also assure that everybody with a need to know *does* know about this project. Stops people from claiming later, "I never knew about this." Not to mention that the more people are involved in the responsibility, the fewer crucifixions are likely if the project goes down the tubes.

To get the data and details needed to fill out the approval form, project managers, team members, and others now must engage in project planning.

16 CHAPTER PROJECT PLANNING

No one can give any one a project of which nothing is known. Knowledge of a project comes from planning it. Project planning methods bring the project from less known to more known, from broad strokes to detail, from macro to micro, from uncertainty to certainty.

PROJECT UNCERTAINTIES

Although the premise behind a project may be true, the project itself may be so stated that we seem to know nothing. "We, uh, better, like, do something about the radiation." But what? Toward which purpose and goal? Projects start full of uncertainties (see Figure 16-1), but *uncertainty yields to definition.*

PROJECT DEFINITION

Projects can be brought from uncertainty to certainty; they can be defined by the use of any of these three methods:

- Structured planning

Projects are full of uncertainties.
Uncertainty of Activity—What work is involved in this project? The activities and tasks that must be done, carried out, executed to bring the project to fruition. *Uncertainty of Sequence*—What work, activity, and task precedes and follows or depends on what other work, activity, or task? *Uncertainty of Cost*—How much will the work, an activity or task, cost in monetary and other resources? *Uncertainty of Time*—What is the execution time of each activity or task? What will be the duration and schedule of the work and project? *Uncertainty of Control*—Considering all of the above, how will we manage the work and project, and how will we monitor its execution?
Uncertainty yields to definition.

FIGURE 16-1. Project Uncertainties

- Deductive or backward planning
- Inductive or forward planning.

Each method works best for its own set of circumstances. In some projects, the method that works best is a combination of these techniques.

Structured Planning

Ultimately all project planning should be structured. Structured planning, however, is called that because it starts with a predetermined structure. "This is the pattern, make the project fit it." Structure may be mandated by the internal methods of the organization or by the usufructuary (or client).

The classic structure is the *work breakdown structure* (WBS) formatted to isolate the details of the project. The structure pushes project planning into defined patterns. Figure 16-2 shows a very common, product-oriented WBS. WBSs are made up of layers called "levels of indenture," a name that comes from the way the earliest WBSs were prepared: on the typewriter. Figure 16-3 shows that classic arrangement for an action-oriented WBS.*

---

*Generic numbering systems, such as shown in the examples, can be used to show how the components in the structure relate to one another. There are, however, any number of systems that can be used for this purpose.

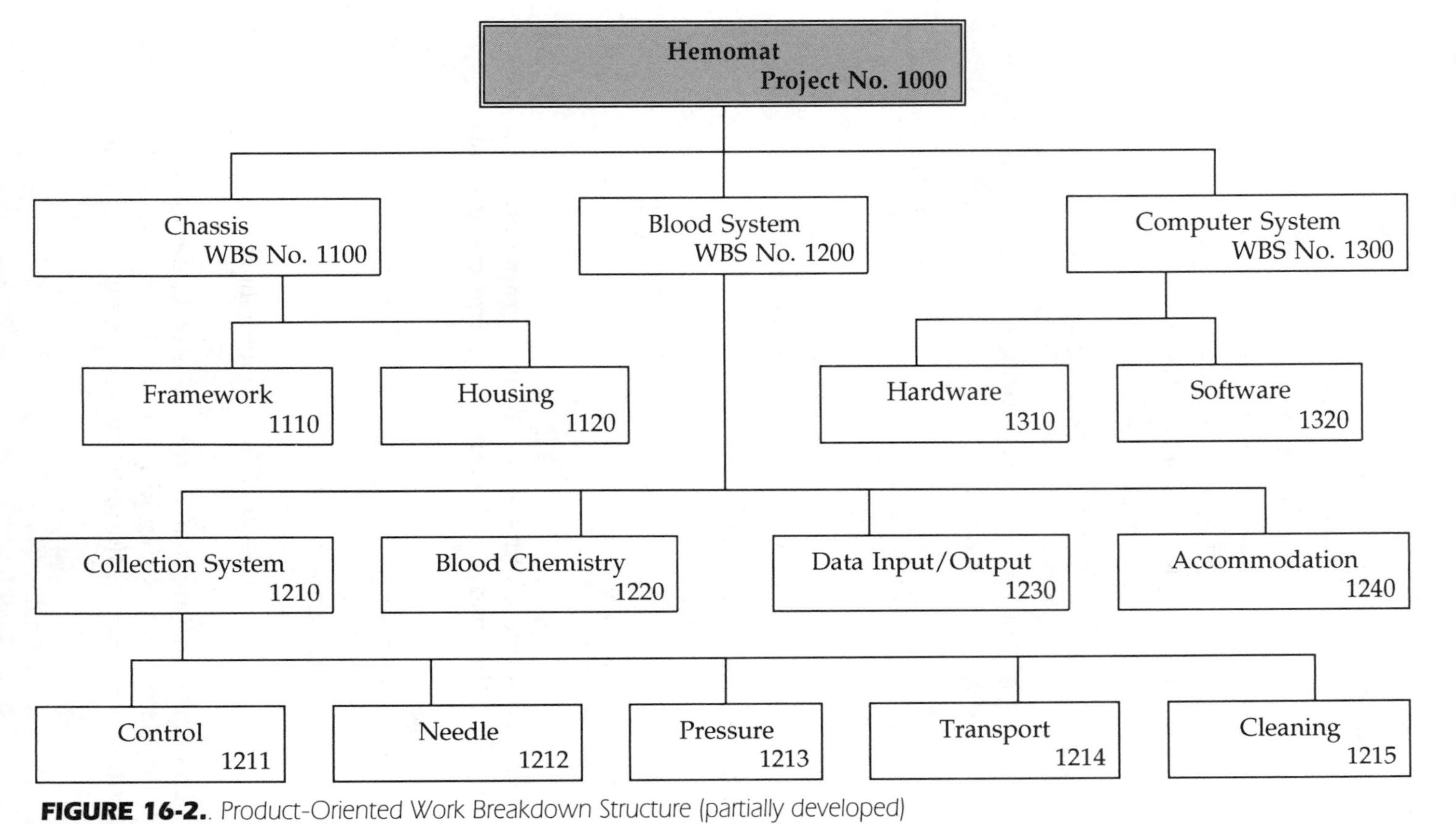

FIGURE 16-2.. Product-Oriented Work Breakdown Structure (partially developed)

From *Practical Project Management*, Anton K. Dekom, New York: Random House, 1987. ©A.K. Dekom. All rights reserved.

- 2000 Hazard Reduction Program
  - 2011 Review existing hazards
    - 201-1010 Review actual cases
      - 201-1020 Frequency and severity
      - 201-1021 Interviews of actual cases
      - 201-1022 Equipment and process cases
    - 201-2010 Indirect and secondary effects
      - 201-2020 Documented cases
    - 201-3010 Statistical extrapolations
  - 2021 Existing laws and regulations
    - 202-1010 OSHA, other federal
    - 202-2010 Local codes
  - . . .
  - 2061 New safety program
    - 206-1010 Training
    - 206-2010 Policy and procedures
      - 206-2018 Distribution and posting

FIGURE 16-3. Classic Work Breakdown Structure

Each vertical line or division is a level of indenture. The numbering system is a variation of a generic system, used to indicate breakdown levels and work packages. Everybody, moreover, can tell that all the components must belong to Project 2000.

No fixed nomenclature exists for the components of a WBS or of a project. The following hierarchy of terms could be used:

- ☞ Project
  - ☞ Key phase
    - ☞ Activity
      - ☞ Work package
        - ☞ Task
          - ☞ Individual contribution

This nomenclature may further be used to define levels of control. Project managers should not have direct control at a level of indenture lower than the activity. (The only exception is if the project has only two people on the team and both work, hierarchically, for the project manager.)

Deductive Planning

Deductive or backward planning is used when the client knows what she wants and expresses it. A customer orders breakfast: two eggs over light, four rashers of bacon, two pieces of toast slathered with butter, orange juice, and coffee to float a battleship. The last activity will be "the breakfast is in front of the customer." Immediately previous, somebody will "set the breakfast in front of the customer," preceded by "take the breakfast to the customer," and so forth.

Data processing applications may be developed through backward planning. "This is what we need: a system to register all people coming and going. A log for their physical data entry; a program to computerize it." The project may start out by defining what the log might be, how the person entering or leaving will identify himself, and what, if any, badging system may be called for. We might get a badge reader, data entry, door control, and the computer program that collects and reports entry and exit data. Each is a job that works itself backward from the end item.

Inductive Planning

Inductive or forward planning is used when the fruit of the project is not known or the client can't quite put it into words. "We don't know what we should do, but let's do something, just to get this off the ground."

Our assignment is Project #4000, "Find a cure for DRTB, drug-resistant tuberculosis." Yes, but how? "Well, why don't we start by having somebody find out what we already know about the disease?" Why? Because we don't want to duplicate effort and because we want to benefit from what is already known.

"Somebody"—a person with the necessary qualifications for this task—will be put in charge of "data collection from existing research." This person must have sufficient degrees and the appropriate personality to interface with various researchers, document thoroughly, and so forth.

One of the more knowledgeable planners might suggest that we should not concentrate on that narrow scope, given the way the disease is spreading, but broaden to a key phase: "Ascertain the etiology (origin and cause) of the disease." This planning person points out, furthermore, that the liaison job might pursue any of four activity streams:

- 4000-4100 Etiology of DRTB
  - 4000-4110 Review literature
  - 4000-4120 Liaison with research centers
  - 4000-4130 Sampling of actual sick people
  - 4000-4140 Search cases of deceased sufferers

The administration-oriented planner, meanwhile, thinks about the back-up this (ultimately vast) project will need and suggests

- 4000-4900 Program management
  - 4000-4910 Documentation system and management
  - 4000-4920 Team reporting, management, and administration
  - 4000-4930 Financial control and reports

Because this is a project deeply regulated by and probably funded by the government, it is recommended by another participant that we have

- 4000-4800 Compliance management
  - 4000-4810 Filings, regulatory compliance, and liaison
  - . . . .
  - 4000-4840 Safety and security procedures
  - 4000-4841 Security system

We are still far away from the fruit of the project, a cure for DRTB. But what we do have, if the above or something like it (with or without all those numbers) is used, is a start and road map for moving ahead with the project. Not to mention that by doing this planning, we can see how the program is expected to develop, what sort of people we will need on the team, and what their tasks might be.

STATE THE TASK

One of the reasons for project planning is to arrive at the task. The task may be called a work package—a piece of project work that can be assigned, described, costed, and scheduled. As shown in Figure 16-4, tasks have at least four things in common—name or designator, timing, cost or budget, and standards of performance or acceptance—together with some related responsibilities.

Responsibility by Type

There are three types of responsibilities for each assignment or task:

- *Responsibility for action*—Do a specific job. Design the engine, interview the patient, write the program; the most common assignable responsibility.
- *Responsibility for control*—Check the finished job. Inspect the part, proofread the interview report, make sure the barge moved far enough.
- *Responsibility for action and control*—You are responsible for doing this job and making sure it's done right.

Responsibility for action is normally given to one person and responsibility for control to another. Action (assembler, for example) is reviewed by control (the quality control inspector). This is the world's favorite way.

There are two times, however, when responsibility for both action and control may be in the same hands. Higher management

All tasks have at least four things in common:	
■ Name or designator	• Task identifier, "handle"
■ Timing	• Execution time
■ Cost or budget	• Costs projected or allowed
■ Standards of performance	• Measuring what it does
or	
Standards of acceptance	• Measuring against requirements
Assign the task	• Tasks must be assigned so they are associated with one person
Assign responsibility	• A manager or contractor (the person, not the company) with responsibility for executing the task and for the worker doing it
Stay in control	• Follow up on the execution and conditions of the task

FIGURE 16-4. Tasks, Features, and Rules

is responsible for both. The division's new general manager is told, "You're in charge. Make sure the stuff you ship is good." Action and control are also in the same hands when the activity is given to a person of unique skill or specialty who really can't be directly checked. "Only you know how to make invisible aluminum and make sure it really is good stuff." The "good" of the stuff, of course, might first have been stated in terms of standards of performance or standards of acceptance.

Responsibility by Degree

We recognize four degrees of responsibility:

- *Direct*—This is **your** job, **you** do it.
- *Supporting*—Help her carry out that job.
- *Monitoring*—Watch how they do it and report.
- *Supplying*—Vendors, contractors, other outsiders.*

Responsibilities must be shown on task and other project documents. Forms for the purpose abound, but one of the simplest can be the typical work order. A project-oriented sample is shown in Figure 16-5. Responsibility can be shown

- With a person's name, badge number, other identifier
- Using a department number or code
- Indicating a vendor or contractor name, code, and/or representative
- Using a skill code (e.g., 411 = mechanical engineer).

Project leaders always should reduce responsibility to one person's name. Nobody is named Electrical Department or Programmer, but we all know Early Schmidt, who is the project accountant responsible for cash flow control.

The rule of naming names should also be followed when dealing with contractors and vendors. Montmorency Clinic has a lot of people working for it, but Dr. Mohammad Zambiq is a specific person; you can even get him on the phone.

Now that we know the tasks, we can schedule and cost and do other things with them. First, however, we must know a few basics of project management, some set-ups.

---

*"Outsider," in this context, is always a vendor or supplier. It can also be another division of the organization, but should never be used to refer to a sister department within the organization.

<table>
<tr><td colspan="3">PROJECT WORK ORDER</td><td colspan="4">Activity or task</td><td>Number</td></tr>
<tr><td colspan="6">Project and key phase</td><td>Number</td><td>W.O. date</td></tr>
<tr><td colspan="8">Description of work</td></tr>
<tr><td>COSTS</td><td>In-house labor
$</td><td>Out-of-house labor
$</td><td>Materials
$</td><td>Contractor
$</td><td colspan="2">Charges
$</td><td>TOTAL
$</td></tr>
<tr><td colspan="3">Person responsible for the work</td><td colspan="4">Department, contractor</td><td>Phone</td></tr>
<tr><td colspan="3">Person accepting the work</td><td colspan="4">Department</td><td>Phone</td></tr>
<tr><td>Work is based on</td><td colspan="6">Drawing / Task assignment / Change order / etc.</td><td>Name, number</td></tr>
<tr><td colspan="8">Others authorized to charge to this work order</td></tr>
<tr><td colspan="4">Control over this work assigned to</td><td colspan="4">Scheduled start / finish dates</td></tr>
<tr><td colspan="3">Work order prepared by</td><td colspan="5">Work order approved by</td></tr>
<tr><td colspan="8">☐ Attachments</td></tr>
</table>

FIGURE 16-5. A Project Work Order Format

17
CHAPTER
PROJECT SET-UPS: THE BASIC TOOLS

Cooking presumes the availability of at least one item: heat. In project management, where the concept of managing implies the need for control, that necessary item is control. For control we need control points: milestones.

MILESTONES

Milestones are not points at which change occurs in projects; change is continuous. Even when nothing is being done, change goes on. Milestones are points at which change is evaluated. Milestones do not exist in nature—no word, concept, or noun is a milestone without being defined as such.

Project managers define milestones in collaboration with those who contribute to the project. The various kinds of milestones are listed in Figure 17-1.

Once we have clearly described the task, some milestones can be easy to define. The fact that some functional managers and others may not wish to define them results from

TIME	*In units:* *As parameters:*	Per hour, per week The schedule
MONEY	*In units:* *As parameters:*	Units per dollar, dollars per unit The budget
EVENT	*When "what" occurs:*	"When the paint is dry" "When you start debugging"
PERCENT	*Ratio between factors*	
HYBRIDS	*Combinations of any or all:*	"Make two, at $12 each, by Wednesday"

FIGURE 17-1. Milestones

- Project planning that is not clear or detailed enough
- Resource providers that may be reluctant to give or agree to milestone facts.

On one hand, numeric milestones are easy to express. There are, after all, people who can estimate the cost of making the movie by looking at the script. On the other hand, some people claim they must be "free" to be creative; others just plain refuse to commit.

EARNED VALUE

Whether people want to agree and contribute or not, all but one of the types of milestone are easy to define. The difficult one is percent. About the only percent we can define with certainty is 0 percent. Maybe. With a lot more difficulty, we could also agree on what is 100 percent. We are back to the principle that milestones must be set by the project manager in agreement with the resource providers.

One of the more useful systems of setting and handling percent milestones has been variously called PMS, Project Monitoring System or Progress Measuring System, or EV, Earned Value.*

*The U.S. government's C/SCC (Cost/Schedule Control Criteria) is one of the systems based on the EV structure discussed here.

Set-Up

EV is a method to establish and use percentages to tie progress to the value of that progress. It is the ultimate hybridization of each method of milestoning. The set-up calls for four families of facts:

- Through project planning we describe a task (e.g., work package).
- Tasks are stated with their cost, time, and event, creating a "value."
- A percentage (of completion) is assigned to each value.
- Comparison of the budgeted achievement of that task to time and cost conditions set for that achievement yields the earned value.

We thus get a statement such as

Task No. 123-456:	Prepare the breadboard model
Budgeted cost:	$37,000
Scheduled finish:	February 29
Percent assigned:	13%
Condition:	Meet specification 123-456

Because percentages are arbitrary, EV processes must have the concurrence of the functional people (or contractors). An EV set-up is detailed in Figure 17-2. Figure 17-3 is a convenient form for arraying tasks and EV data. The form's reference to meetings is to

Break the project into tasks, work packages, activities.

Set the budget for the task.

Set the time for the task.

State the standards of performance or acceptance.

Establish a percentage for each task on the basis of:

- Importance to the project
- Importance to the usufructuary
- Importance to you
- Effect of the task on project progress
- Cost of the task in relation to the total budget
- Time relationship to the project
- Your gut feeling

Negotiate, but start out with a plan and know what you want.

FIGURE 17-2. Getting EV Started

EV ASSIGNMENT FORM	Project		Number
Prepared by		Phone	Date
Assignment based on meeting of (participants):			Meeting date

Line no.	Activity, task, work package	Number	Budget $	Finish date	% assign.	Reference for accept. stds.
		Totals				

Meeting minutes reference

Distribution ☐ If continued ☐ If continuation

Page of pages

FIGURE 17-3. Defining PMS Values

show when and by whom agreements were reached, since agreements are usually worked out in formal meetings.

To make the process work, the project manager must

- Know why a percentage or range should go with a specific task

- Negotiate with internal and external contributors to the agreement.

All percentages together must add up to 100.

EV in Operation

To operate EV we must translate the data in the Assignment Form (Figure 17-3) into the three sets of values shown and defined in Figure 17-4: BCWS, BCWP, and ACWP. "Earned value" will come from the comparison among those values.

Finally, we must know the sources of the data. BCWS, the Budgeted Cost of Work Scheduled (and the information needed to know it), comes from the Assignment Form. Note that BCWS is also a planning milestone.

BCWP, the Budgeted Cost of Work Performed, is based on our data—the budget—and is compared with data reported by those

BCWS Budgeted Cost of Work Scheduled

Budgeted cost of the work scheduled for this period:
Fill one bag of sand by time T_x for $1.00.

BCWP Budgeted Cost of Work Performed

Budgeted cost of the work actually done in this period (or by time T_x):
One bag of sand filled by time T_x for $1.00.

ACWP Actual Cost of Work Performed

Actual cost of the work done:
Filling one bag of sand cost (over, under, exactly) $1.00.

Value achieved can appear in any number of varieties:

a.	1 bag filled by time T_1 at $1.00	= 100% EV
b.	1 bag filled by time T_1 at $2.00	= On schedule, cost overrun
c.	1 bag filled by time T_1 at $0.50	= On schedule, cost saving
d.	2 bags filled by time T_1 at $1.00 each	= Ahead of schedule, on cost
e.	2 bags filled by time T_1 at $1.25 each	= Ahead of schedule, cost overrun
f.	1 bag filled by time T_2 at $1.00	= 50% EV

Earned value is expected achievement (BCWS) *versus* actual achievement (ACWP) within a time limit:

BCWP – BCWS = Schedule variance, if different
BCWP – ACWP = Cost variance, if different
BCWS – ACWP = 100% EV, if not different

FIGURE 17-4. What EV Will Tell You

doing the job: "We did job J, budgeted at $B, to be finished by F, under condition C."

Comparison. When we compare the work done, at whatever cost, against the work planned at budgeted cost, we get the earned value.

BCWS: 1 bag of sand, by time T, at cost of $1 = 100% of sandbag.
BCWP: 1 bag of sand, by time T, at cost of $2 = 100% of sandbag.

This work was done on time, but over budget. EV = $1 and 100 percent. The problem is how to handle the cost overrun.

BCWS: 1 bag of sand, by time T, at cost of $1 = 100% of sandbag.
ACWP: $0.50. Report indicates we have filled only half a bag of sand.

This work is done on budget, but behind schedule. EV = $0.50 and (maybe) 50 percent. The problem is how to handle the behind-schedule condition.

BCWS: 1 bag of sand, by time T, at cost of $1 = 100% of sandbag.
ACWP: $2. Report indicates we have filled 2 bags of sand.
BCWP: $2. We are ahead of schedule and have overrun the cash flow forecast, but are not over budget. 200 percent of sandbag.

To reduce this to a formula:

BCWP–BCWS If there is a difference between the two numbers, there is a schedule variance.
BCWP–ACWP If there is a difference between the two numbers, there is a cost variance.

When the actual work, time, and cost equal the budgeted work, time, and cost, then EV = 100. EV varies up or down with the difference.

Percentages. One additional benefit comes from the use of an EV structure: percentages. Since each task or work package also carries with it a stated percentage, when that job is done, the percentage is achieved.

Earlier we set up Task No. 123-456, "Prepare breadboard model," to which 13 percent of the project was assigned. The task is finished, so 13 percent of the project is complete. Suppose that, somehow, this task doesn't get finished. The project cannot be more than 87 percent complete. With the EV system we always know why a certain percentage of the job is finished—or not finished.

When Task 123-456 is finished, the contributing department reports that this task is 100 percent finished, and the project manager reports that the entire project is 13 percent finished. Using this example, the EV system sets

13	=	100
Task 123-456's portion of the finished project	=	the portion of Task 123-456 that is complete

Not to mention that if Task 123-456 is partially finished, and the finished segment can be expressed in percent, we can still measure the portion of the whole project that is complete.

Retention

Having earned a value of 100 percent does not mean a 100 percent payout! Consider "retention": although the contributor has, in fact, earned 100 percent value, the immediate payout will be less by the amount of retention.

Say that retention is 20 percent. Although the task has EV = 100, only 80 percent of the payment due is disbursed; 20 percent is retained until some future moment. The future moment may be tied to an event: when the paint dries, when the new program is incorporated into the procedure library, and so forth. Or the retained amount may be withheld until the end of the project, or until the supplier proves that they paid their suppliers.

CADENCE

As with every other type of milestone, time must be defined. The time division by which the project is planned, executed, and controlled is called the "cadence." Cadences are the common time divisions: days, weeks, months, and the like (see Figure 17-5).

Project managers set the cadences on the basis of these criteria:

- Scheduled duration of the project
  - A 7-month project should not be cadenced in months; a month is too long for effective project monitoring.
  - A 7-year project should not be cadenced in days; monitoring that frequently might drive everyone crazy.
- Frequency of change in the project
  - If significant changes are frequent, the cadence might be tight to allow frequent monitoring.
  - For project times when little happens, the cadence might be looser.
- The client
  - A nervous client necessitates a tight or frequent cadence.
  - A loose client, however, doesn't mean a too-loose cadence.
- The project manager's own control needs
  - Familiarity (or lack thereof) with the work or with contributors will influence the cadence chosen.

Cadence is shown either in the upper right hand corner of project documents or in a legend block. As with everything else, cadence has to be negotiated by the project leader with those affected. Additional rules regarding cadence are in Figure 17-6. The last of these rules—*Lay the plan on the secular calendar.*—will be most important in project scheduling.

Calendar day	=	÷ 365
Working day	=	÷ 260
Week	=	÷ 52
Month	=	÷ 12
Quarter	=	÷ 4
Trimester	=	÷ 3
Semester	=	÷ 2
Year	=	÷ 1

FIGURE 17-5.
Cadences for Projects

Set the cadence: It would be nice of upper management or the usufructuary would do this, but it usually falls to the project manager. Know what you want, but be ready to negotiate.

Same cadence per level of indenture: Contributors working on one level of indenture should have the same cadence. Be certain that all the people who meet at a specific date have the same cadence. This tends to assure that they do, indeed, meet on that date.

Mature the cadence: Alter the cadence as the project moves forward. Watch for:

- Brick walls—"It didn't work out as planned"—Ease the cadence.
- Breakthroughs—Unexpected (unscheduled) serendipity—Quicken the cadence.

Lay the plan on the secular calendar: See what the real calendar—with seasons, weekends, and holidays—does to your schedule.

FIGURE 17-6. Rules for Cadencing Projects

CHAPTER 18

THE SCHEDULING PROCESS

Scheduling is a process; it is not an arithmetic operation, although arithmetic is a necessary part of it. For example, say we have two tasks. Task 123 has an execution time of 5 hours, and Task 234 has an execution time of 6 hours. The second task follows the first. When will these two jobs finish? We don't know.

EXECUTION TIME

What we do know, with the aid of arithmetic, is that the total execution time of these two tasks is 11 hours. Execution time is the basis of all scheduling. Execution time asks the question, *How much work is there?* Answer: 11 hours' worth.

To get to execution time, we must—as always—start with the task because the task is (or should be) small enough that its execution time can be determined with some certainty. This statement further implies, correctly, that the smaller the task, the more easily its execution time can be determined. As stated earlier, uncertainty yields to definition.

restricted ways of doing the job—only a proof printer can operate the proof press, even if only one proof is needed.

Necessity also comes into play: it's freezing and we're working outdoors. It's hot, raining, or there are other conditions that slow the job down. Simple things, like letting a mix cool, for example, or heating it up. And always, we must factor in the effects of the secular calendar.

THE SECULAR CALENDAR

A key modifier of the schedule is the secular calendar. It has both macro and micro effects. Macro effects consist of such things as the seasons and aspects—besides the weather—related to the seasons: hunting season, the World Series, the running of the smelt, holidays, vacations, that fiddling contest Maria never misses, and so forth.

Micro effects operate on the day or the week. Some people's first act of office is to get coffee or stop at the bathroom. There is the slow-down toward lunch and go-home time, starting late on Monday and stopping early on Friday, coming back from a long weekend, "glad to be back at work so I can get some rest."

One can overdo these considerations by getting too detailed. When necessary, the best course is to come up with an efficiency quotient and apply it to the hours or total schedule. How to do this is explained below.

When the duration is established by whatever process, the effects of the secular calendar must be worked into duration. Three conditions will affect how it all works out:

- The job of doing all this is long and tedious; you might not have enough time and even less inclination to do it.
- The new truth emerging from working the process correctly might not be acceptable or desired by others.
- "Drop dead" deadlines—those imposed by outside conditions—might negate all this fancy footwork.

One thing you can do is work these conditions into an efficiency value (EV). What is the relative efficiency at which the people in this organization work? You say 80 percent? Well, modify execution time by the efficiency quotient and use the resulting product. Assume that the job has an otherwise calculated duration of 60 hours. At 80 percent efficiency, that gives an expected duration of 75 hours:

$$\frac{\text{Duration}}{\text{Efficiency}} \times 100 = \text{Expected Duration} \;☞\; \frac{60 \text{ hours}}{80\ \%} \times 100 = 75 \text{ hours}$$

DURATION

Duration asks the question, *How much time will it take?* With all the factors summarized in Figure 18-1, we add and modify by calendar and efficiency factors and come up with a duration. This is the amount of time that passes, not the sum of working hours. That sum we will need for costing, covered in the next chapter.

- ☞ **Execution Time** *How much work is there?*
  - ☞ **Common Time**
    - ☞ **Laws, Regulations**
      - ☞ **Union Rules**
        - ☞ **Necessity**
          - ☞ **Secular Calendar** *Macro ... micro*
            - ☞ **Efficiency, Wastage**
              - ☞ **Culture** *Within the industry ... within the organization*

When combined, these result in:

☞ **DURATION** *How much time will it take?*

FIGURE 18-1. From Execution Time to Schedule

tion, set up, and definition of a cost item is normally done through a chart of accounts, a section of which is shown in Figure 19-1. Charts of accounts should tell

- What type of cost is it?
- How does the entity define this cost?
- What classification number is given to this cost?

The chart may also tell what stream of subsidiary costs makes up a "main" cost item. For example, account #100, Labor, may (or may not) be made of the following components:

- Actual wages paid our own workers
- Straight time or overtime
- Premium time or overtime
- Cost of benefits (some or all)
- Contract labor.

Some of these subsidiary costs may be split among several main accounts, and main accounts will consist of several subsidiary accounts. Project managers should research and know them, if for

MANAGEMENT INFORMATION AND REPORTING SYSTEMS	
Title: Responsibility Area Expense Accounts and Descriptions	**Procedure no.** 102
Applies to: Corporate Operations and Process Division	Appendix I **Page** 7 **of** 7

Responsibility account no.	Description
920	Interest expense – Interest expense on borrowed funds, such as interest on notes, mortgages, and bonds.
990	Computing service – Transfers of costs to projects by computer operations for computer services.
993	Reproduction – Transfers of costs to projects by the reproduction department for reproduction work.
994	Analytical service – Transfers of costs to projects by laboratories for laboratory analyses.

FIGURE 19-1. Chart of Accounts (section)

Note: In this organization, charges for computer, reproduction, and analytical services are allocated to projects. Also, Accounts 990, 993, and 994 are used for transferring expenses between projects. These expenses appear only on project reports, never on responsibility reports.

no other reason than to avoid being surprised by month-end cost reports.

For what we need now, we proceed to cost classifications—of which there are many and many ways of classifying. A listing is included as Figure 19-2.

Labor Material Other	Direct Indirect Overhead (general, administrative, burden)
Commitments Accruals Reserves Payments	Charges Allocations
And others, as the organization's accounting may dictate.	

FIGURE 19-2. Typical Classifications for Project Costs

Basic Classification

Project managers are normally responsible for allocating or budgeting basic costs, those for which they can be held responsible. Basic costs are

Labor—The cost of employing, using, having humans at work, available to work, or actually working on a project.

Material—The cost of materials, parts, components, assemblies, and other items procured for the project. These items can be procured from an outside vendor, contractor, or supplier, or issued from existing inventories.

Other—Costs that are neither labor nor material, but are incurred for the project in one way or another, or charged as part of running the project, or allocated because the project has to share the burden of other entity costs.

We will encounter more "Other" as the chapter goes on.

Direction

Costs can be further classified:

Direct—Visible costs that are or can be associated with a task; a purchase or expense that can be directly traced to the project, one of its activities, tasks, or needs. Examples:

- A research scientist spends time mixing a compound for a project task. The scientist's time is classified direct labor. Materials used to make the mixture are classified direct material. If the mixer is rented for the task, its cost may be classified direct other (subclass, rentals and leases).
- A ditch is being dug for the project. Ditch digging is direct labor (no material, for the moment). If the digging machine is rented for the task, cost may be direct other (subclass, rentals and leases).

Indirect—A cost attributable to the project, but not visible or not directly traceable to one task or job. Examples:

- The mixed compound (from above) shares space in a cycle oven, where it stays for two weeks of hot and cold. This cost is classified indirect other.
- A lab inspector checks all mixtures daily and annotates the control tags of each mixture. This cost is indirect labor.
- The inspector uses a reactive strip to test each mixture and throws away the strip after review and record annotation. The strip will probably become indirect material.

Other—Other costs, neither labor nor material, that somehow are incurred or increased because of the project. Their apportionment to the project may be visible and justifiable only indirectly. Examples:

- A chemical toilet on a construction site.
- Beefing up the guard service because of the project.
- Using the computer processing stream to debug the program being written for the project.

Each expenditure has to conform to the chart of accounts and coded so that it is correctly and consistently charged throughout the project by everyone.

Overhead

In our structure, overhead (O/H) is a part of "other." Also called "burden," overhead is a way of charging, of burdening the project with costs, parts of which can be shown to be part of the project in traceable ways. The president's salary is an example. A part of the president's salary cost should definitely find its way into the cost of the project, especially if the project is for an outside client. Overhead comes in layers.

Overhead is not just one item; it is a collection of accounts and items. Important to the project leader are two forms of overhead, engineering O/H and manufacturing O/H.

Engineering O/H. All O/H is a cost that, somehow, is apportioned to other accounts. Engineering work is overhead in most cases. In the work of design and facility engineers, for example, their support people and other costs incurred by them are typically part of Engineering O/H. Parts of these, and any number of other similar costs, will be allocated to the project by whatever formula is devised.

In an engineering firm, however, these same costs may be direct costs to the project. That is because engineering is the business of the firm. Architects' work is often part of Engineering O/H. In an engineering and architectural firm, however, the architects and their cost may be direct.

Manufacturing O/H. A small group of industrial engineers is attached to the office of the Director of Manufacturing. These people may conduct time studies, coordinate plant relocations, and help with equipment specifications. The costs associated with their presence on the payroll and their work will most likely be considered Manufacturing O/H.

Their work, the work of production engineers and similar others, is performed for the benefit, improvement, or control of manufacturing. Nor can their cost validly or normally be shown against an individual product. Thus they—and any number of other costs—become Manufacturing O/H.

General and Administrative O/H. G&A may contain the costs of general taxes paid by the organization, the Accounting Department, the telephone operator, the president, other overhead costs (e.g., advertising). G&A is normally the last or top layer of overhead with which projects and other working parts of the entity are burdened.

Burden. All overhead is burden. Normally, however, what is actually called "burden" is a layer of overhead closer to the project than the president's salary. Burden is applied to the cost of manufacturing and to projects. Burden may be called "indirect overhead" and consist of the cost of the Plant Manager, the plant staff, production scheduling, maybe all or some costs of an MIS (management information system) function, and so forth.

Within certain limits, local accounting policies govern. Project managers, who are responsible for cost management, must therefore learn the local policies, including what is direct and what is indirect.

Indirect and Direct. Overhead can be direct or indirect. The salary of the general foreman in a manufacturing plant can, depending on the rules of the game, be indirect cost, indirect overhead, or direct overhead.

Most often, the general foreman is direct overhead because

- A higher ranking foreman is needed to supervise a major segment of production and several direct foremen.
- The cost of the higher ranking foreman must be included in each item produced.
- Which part of that foreman's pay should be charged to the item now in front of a direct worker is not discernable.

Therefore, we charge the cost of the general foreman to all items produced and whatever else is made under the overall control of that general foreman.

Charging the factory superintendent's cost is not easy at all. Under that superintendent might be the production lines of a number of products. Not to mention that within the cost package we call "superintendent" may be a process engineering group, production scheduling and control, stock clerks, certain taxes, and so forth. Those become indirect overhead.

Charges

Overheads and burdens are used to (well...) burden the direct costs usually in one of two ways, by allocating or by charging. Which way is chosen generally depends on company policy. "If that's the way they want to do it...," as long as the costs are shown and properly accounted for. The real reason for proper accounting, in any country in the world, is tax law.

Allocating. When an organization allocates burden or overhead, that means the costs are spread in layers over the direct costs, usually on the basis of a formula. Most often, the formula is a multiplier that operates on direct labor.

Say the burden amount is $800,000. This burden is to be allocated over $400,000 production cost (whatever that may mean), and $200,000 of that production cost is labor. In this example, every dollar of labor is burdened by $4 of overhead: overhead is 400 percent of direct labor.

The project manager is and should be responsible for direct labor dollars. Burden is something else. If the burden rate really were 400 percent, this job would cost $1,200,000 by the time the burden is allocated. That can be a surprise.

On top of the burden, other costs may be allocated: for MIS, imputed interest for tied up working capital, corporate staff support, and general maintenance. Specific maintenance and requested MIS services, for example, may become "charges."

Charging. Costs incurred by or for the project are charged to the project. Equipment used on the project breaks down and has to be repaired. Maintenance gets a work order, signed or countersigned by the project leader: "Please repair. Charge to Project 2000." The cost of that work order (yes, burden may be added to the cost) will be charged to the cost of the project.

Or, thanks to the fact that the project calls for a whole flock of new reports—this is a large project—MIS is being asked to produce them. MIS charges $1,200 per month. This is a charge in fact, because it is due to and can be traced to the needs of the project.

Unfortunately, this is not the end of the story. By the time this charge is fully "costed out" by Accounting, the project may be charged a total of $2,800. Burden, of course.

THE REST

The rest may affect only the cash flow. First are commitments.

Commitments. The project manager has agreed with a vendor to procure 5,000 bags of sand at a delivered cost of $5,000. Unless the project manager has overstepped the bounds of authority and the vendor knows it, the organization is now committed to a lot of sand. The purchase order will be issued on Monday.

Commitments are promises to pay (on the vendor's side, of course, promises to deliver or perform). Until they are documented, commitments may be invisible and surprise the organization if the invoice is not expected.

Accruals. An accrual can be real or only on the books (to become real in some form or another, sooner or later). An accrual occurs when a future expense is expected and smaller sums of money are set aside or accrued to cover that expense. Take an employee who gets paid weekly, but pays the rent or the mortgage monthly. That person may have an accrual system. The paycheck is paid directly to the employee's bank account. On that employee's personal budget book, a fraction of the paycheck is accrued for rent or mortgage.

In reality, all of the pay is in the bank, not even separated from other people's money! The employee has, however, accrued half of the rent or mortgage payment by the second week of the month.

In projects, too, future cost may be accrued or "accrued against." Accruals may have their main influence on cash flow. In other words, they could tie up cash in advance of their being needed for the actual payment.

Reserves. A reserve is a form of accrual against a future actual or potential event. Your company may keep a reserve for overtime, just in case. Or for price increases, or even for the defaults of others—just in case. Until the event occurs, the reserve may be an actual amount of cash or a bookkeeping entry, "reserved" against that future event.

Both accruals and reserves can be reversed. The amount accrued or reserved can be returned or cancelled or, in proper terminology, "reversed."

Payments. Payments occur when monetary values change hands. Payments can be real, when actual money (cash, check, etc.) is paid out to cover an invoice, a paycheck, the accrual, or the reserves mentioned earlier. Payments can also be by book entry transfers, in which case people talk about "funny money." To the project leaders who get charged for the cost the book entry represents, there is nothing funny about it. Let's get back to the MIS reports, for which the project is charged $2,800. If the charge is from an internal MIS department to the project, no actual cash changes hands. The cost is a non-cash charge, hence funny money. To the project leader, however, the charge is real, since the project will be shown to have incurred that $2,800 cost.

CONCLUSION

These are the foundations, the bare bones needed for budgeting. Principles shown are modified by accounting and other policies of the organization, by tax laws, and by generally accepted accounting practices.

To sum up:

- Costs are direct—those under the control (or presumed to be under the control) of the project manager and the managers of internal and external contributors.
- Costs are indirect—even if some of them are under the control of those managers.
- Other costs may be allocated or charged to the project, by formula or by other methods.
- Basic costs are labor, material, and other (not normally labor or material) in varying degrees under the control of the project manager.

All these costs can be augmented, burdened, and charges added, to absorb other costs of the organization.

It is preferred that project budgets for which the project manager is responsible be made up only of "controllable" costs. All other cost layers should be applied by the accounting department.

Finally, in preparing the project budget, project managers must not only know these basics, they must also know the task. As with scheduling and team assignment, budgeting starts with the task.

20 CHAPTER PROJECT BUDGETS

By now, budgeting should be a series of arithmetic functions, chiefly addition and some multiplication. As with project planning we can use backward, forward, and structured budgeting, or any mixture of these types. We do, however, have to know the tasks, activities, and components that make up the project.

ADDITION

No matter what the method of budgeting, we need to know the task and cost elements. For the purpose of identifying activities, tasks, and their costs, any number of formats can be used, including the one shown in Figure 20-1. This format is normally the detail to support summary entries.

Task Entry Form

This form is largely self-explanatory, except for a few items:

ASAP—As soon as possible. This task needs the earliest possible start (or finish, if Finish also is ASAP).

TASK ENTRY FORM	Project name	Project no.	Date
Key phase, activity, task, work package name			Number
Assembles into ☐ key phase ☐ activity ☐ task ☐ work package			Number
Is part of ☐ key phase ☐ activity ☐ task ☐ work package			Number
Entry created by (name, title, department)			Phone

Description

Work detail	Labor	Material	Other*	Total
TOTAL COSTS				

Start is ☐ ASAP ☐ ALAP	Scheduled for date	☐ As calculated ☐ Range of dates shown
Finish is ☐ ASAP ☐ ALAP	Scheduled for date	☐ As calculated ☐ Range of dates shown
Cadence	Duration	Person responsible for task (name, department, phone)

*If contractor is involved, identify (name, code, contracted cost)	
Method of acceptance (cite, reference)	
Person accepting task (name, department, phone)	
Reference change orders (if any)	Latest change no.

FIGURE 20-1. Entering the Task

ALAP—As late as possible. Start (or end) this task as late as possible. Helpful where quick or many changes can be expected, or where today's model may be obsolete tomorrow.

As calculated—As calculated with a PERT, CPM, Gantt, or other scheduling system, or by the computer.

Range of dates—Used when the task may start and finish between a range of dates, as when float exists or slack has been allocated to the task.

Cadence—Indicate whether the task is measured in days, weeks, months, and so forth.

Duration—The number of cadence periods this task is expected to use.

What the form does not contain, as far as costs are concerned, are Chart of Account numbers. We may decide to add these to ensure that the costs are coded and accounted correctly. The form may also be integrated with a PERT/CPM chart to get the task start and finish numbers ("I" and "J" in PERT notation) or activity numbers (in CPM notation) from those systems.

Operation

After enough copies of the form and adequate instruction have been given to functional and other managers, people are supposed to fill out the forms related to their work. It is neither smart nor desirable for the project manager to fill out all the forms. Data entry and form upkeep should be the responsibility of those people whose task and cost the form represents. Forms should be prepared and data entered by the lowest appropriate level of indenture.

Each cost—labor, material, other—should also be shown on a separate form. Cost data may come from actual calculations, rate books, past organizational experience, price manuals, or other informative sources. Descriptions and costs should be validated by a functional manager, not only to give a sanity check to the data, but also to absorb the necessary responsibility.

Labor Cost

For costing labor, a cost sheet like that shown in Figure 20-2 can be used.* Earlier, the team needed for a task was established. Recently we got additional information:

* It does not matter if the project is monitored by computer. The computer "screen" is, for our purposes, just another form, except that the information *may* be input directly to the computer and not to a paper form. Not to mention that people still have to be trained in handling input (and later, in understanding output), as well as actually inputting the information. Nothing changes, except the medium and the fact that the computer can do the arithmetic that we would otherwise have to do.

PROJECT ASSIGNMENT, COST DETAIL

Project name: Hemomat | Project no.: 1000 | Date: 10/10/98

Task name, detail: Needle, design to breadboard model | Form completed by: Maria T. Celena | Task no.: 1212

Line no.	Employee Number	Employee Name	Dept. no.	Skill code	Hours Planned	Hours Paid	Rate $	Labor cost	Start date	Finish date
1	3456	Arnie Black	DESE	486	400		30.00	12,000	11/11/98	05/05/99
2	4567	Seymour Green	PURC	202	100		25.00	2,500	12/06/98	05/05/99
3	5678	Neva B. Pinkus	MNFG	508	200		25.00	5,000	02/02/99	05/05/99
4	6789	Barry Dresser	QUAL	666	100		25.00	2,500	12/15/98	05/05/99
L2-1		Manufacturing labor cost			1,800			30,000		
L2-2		Design, programming labor			300			8,000		
L5								60,000		
L6								15,000		
L7								75,000		
L8		Contract labor cost						15,000		
L00								90,000		
☐ If continued ☐ If continuation		Page 1 of 1 pages	Totals		2,900			90,000		

FIGURE 20-2. Defining Labor Costs

- Name of the team member
- Clock (or employee) number
- Skill code.

Now, the following data are added:

- Charge rate
- Planned hours (in this example; it could also be weeks or months) or paid hours for which the charge rate is the multiplier
- Start and stop dates—the period of time within which the hours of work have to be rendered.

Bulk Labor. There is no specific space on this form for undifferentiated labor. This refers to labor costs for groups: fabrication, assembly, programmers, scientists, and others who contribute their work to the project, but aren't project team members and are charged as a block or group. For these, we need to set up separate line item costs.

Line	*Item*	*Hours**	*Cost*
L9	Cost of programming	600	$24,000
L14	Assembly	300	$18,000

This gives the discrete cost of specific team members as individual line items. Group or undifferentiated costs are given as a lump sum total for the group or pool shown as a line item. It is important that subsidiary documents show how these costs were determined; this may be necessary to support capital requests or to resolve subsequent problems.

Footing and Summarizing. Cost columns are then footed (added) and new, summary line items created. Figure 20-3 lists summary line items. Because consistency in budgeting is important, establish a rule that each item of a type gets the same line number. For example, Line 00 might always be the summary line:

L00 = All labor costs
M00 = All material costs
O00 = All other costs

*Or whatever increment in which project costs are accounted.

L1	Direct, variable cost of team member labor (under our direct control and responsibility).
L2	Direct cost of undifferentiated labor.
L3	Straight-time cost of overtime.
L4	Premium time cost of overtime.
L5	All internal, direct labor costs.
L6	Benefits, payroll taxes, and other payroll costs.
L7	Total cost of all internal labor.
L8	Contract labor cost.
L00	All labor costs to the project.
M1	Direct cost of material and parts.
M3	Costs of freight in, taxes, customs duties.
M4	Procurement costs.
M7	Total cost of procurement: invoice cost, taxes, purchasing department cost, receiving inspection.
M8	Cost of inventory.
M00	All material costs to the project.

FIGURE 20-3. Summary Line Items

NOTE: Not every line will have full correspondence in every category, however, the major line items will. This discipline makes it easier to know what real costs are in the project and where they come from.

When cost data are entered in their proper place, the columns are footed down (added vertically) and the rows footed across (added horizontally). If the project manager has seen to it that everybody uses the same format for the same things, we can train the eye to look at one rubric and find the answer.

Next, we have to look for summary or higher level formats that integrate the activities, timing, and cost of the project so that budgeting can proceed and the project entered in the books.

The Comprehensive, All-Inclusive, Summary Form

The Project Summary Form, Figure 20-4, accomplishes three critical things:

- Integrates all the factors
- Connects all activities, tasks, and so forth
- Shows how successive totals are arrived at.

PROJECT DEFINITION SUMMARY FORM	Activity no.	Project no.	
Project, breakdown item name	Major components	Timing	
	1.	Sched. start	Sched. finish
	2.	Early start	Late start
	3.	Early finish	Late finish
	4.	Duration	Slack
WBS no. I-J No.	5.	Paid time	Actual time

Responsibilities				Activity description		Resource allocation			
Direct	Suprtg.	Montrg.	Supplg.	Line no.	Detail	Labor	Matls.	Other	Total cost

First meeting	Last meeting	Issue date	Effective date	Change no.	Effective date	Page of pages
Prepared by	Management approval	Usufructuary approval		Contributor approval		Contributor approval

FIGURE 20-4. Project Summary Form

Computerized project management systems still need these inputs, but the linking, the catenation, will be done by the computer program. So from this point on, using a computer can save significant labor costs.

Form Arrangement. The Project Summary Form is composed of five major sections:

Identification—The job and its major components. In the upper left corner, describe the project or key item. At right, under Major Component, tell what components constitute the key item. Every component must belong someplace. There can be a key item without components, but never a component item with no key item (except the project itself).

Timing—Scheduled start and scheduled finish of the project or key item. No resources should be chargeable to the activity outside this scheduled block of time.

Activity—A detailed description of the project or key item. Contains working details only to the extent necessary to let the reader know what is expected from the key item. Space in this section can also be used for:

- References to corporate approvals, procedures, and so forth
- Appendices and attachments
- Internal and trade standards
- Contracts, sales orders
- Laws and regulatory conditions.

Responsibilities—Who is responsible for what aspect or detail, and the degree of responsibility.

Cost—Cost of each line item, by type. The total cost of each key item is a line item in the next higher level form.

THE BUDGET

Budgeting can move from the top down. Through a preliminary project request, assume $3,300,000 has been approved for this project. This initial budget would be distributed to the first level down from the top of the work breakdown structure, the major subsystem level. Each of the responsible entities at the subsystem level would break it down further; i.e., to the sub-subsystem level.

Budgeting can also move from the bottom up, starting with three fundamental points:

- How much will this activity cost?
- What are its major components? What are their costs?
- Foot-up and foot-cross the costs.

Enter all the lowest level work packages into a total for the next level up and repeat until the entire project is completed. In the preparation of an estimate or proposal, this is often done when you assign parts of a project to different individuals or groups, and the proposal manager accumulates costs in accordance with the flow upward through the newly developed work breakdown structure.

Using the Hemomat, shown as a work breakdown structure in Figure 16-2, we reconstruct the total cost (not including burden, overhead, or profit) in Table 20-1.

TOWARD AN INTEGRATED SYSTEM

The generational system of costing is shown in the successive forms of Figure 20-5. The top level (or total project) form—"Hemomat"—should contain all the costs of each activity, task, or compo-

TABLE 20-1.
Sample Total Costs

	Labor	Materials	Other	Total
Hemomat	1,000,000	700,000	1,600,000	3,300,000
Chassis	50,000	50,000	50,000	150,000
Blood system	450,000	400,000	800,000	1,650,000
Computer system	500,000	250,000	750,000	1,500,000
Total	1,000,000	700,000	1,600,000	3,300,000
Blood system	450,000	400,000	800,000	1,650,000
Collection system	50,000	100,000	150,000	300,000
Blood chemistry	200,000	150,000	350,000	700,000
Data input/output	50,000	50,000	100,000	200,000
Accommodation	150,000	100,000	200,000	450,000
Total	450,000	400,000	800,000	1,650,000

NOTE: The numbers are entirely fictitious.

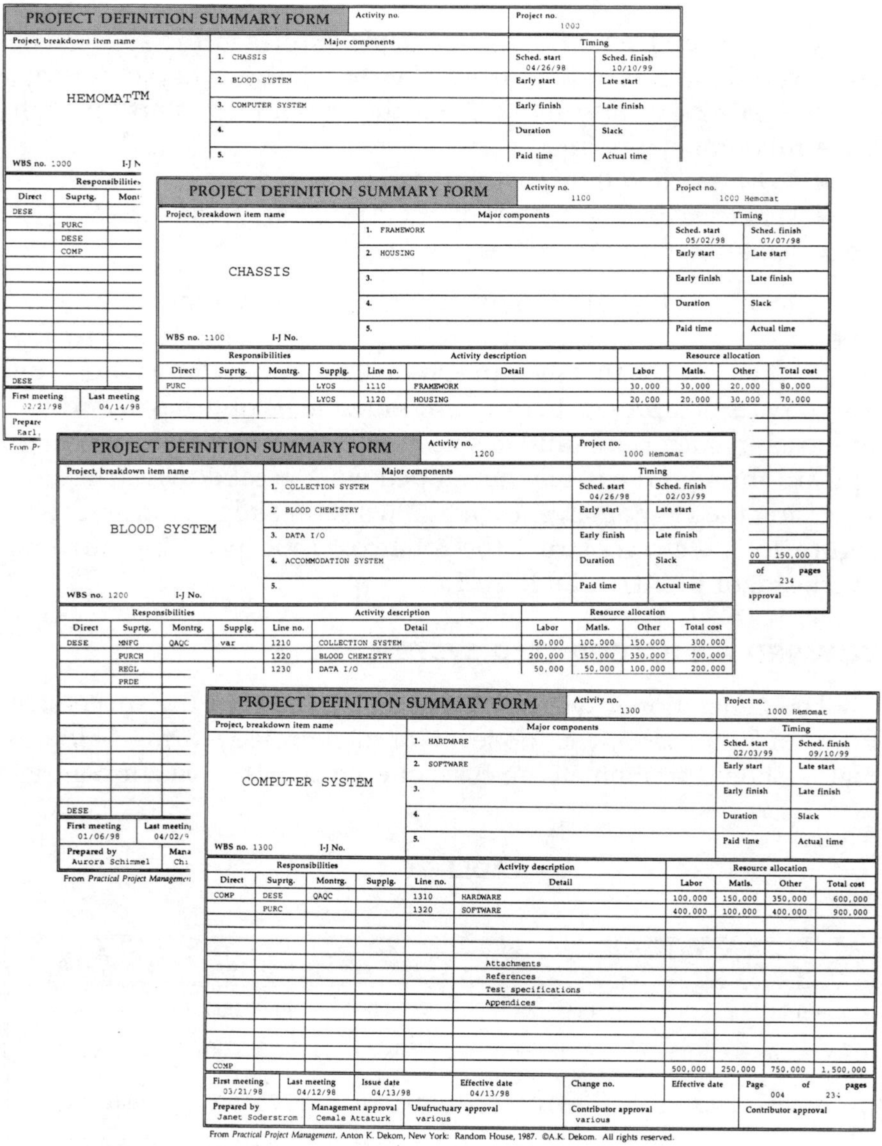

PROJECT DEFINITION SUMMARY FORM | Activity no. | Project no. 1000

Project, breakdown item name	Major components	Timing	
HEMOMAT™	1. CHASSIS	Sched. start 04/26/98	Sched. finish 10/10/99
	2. BLOOD SYSTEM	Early start	Late start
	3. COMPUTER SYSTEM	Early finish	Late finish
	4.	Duration	Slack
WBS no. 1000 I-J N	5.	Paid time	Actual time

Responsibilities		
Direct	Suprtg.	Mont
DESE		
	PURC	
	DESE	
	COMP	
DESE		

First meeting 02/21/98 | Last meeting 04/14/98

Prepare Earl.

From Pr

PROJECT DEFINITION SUMMARY FORM | Activity no. 1100 | Project no. 1000 Hemomat

Project, breakdown item name	Major components	Timing	
CHASSIS	1. FRAMEWORK	Sched. start 05/02/98	Sched. finish 07/07/98
	2. HOUSING	Early start	Late start
	3.	Early finish	Late finish
	4.	Duration	Slack
WBS no. 1100 I-J No.	5.	Paid time	Actual time

Responsibilities				Activity description		Resource allocation			
Direct	Suprtg.	Montrg.	Supplg.	Line no.	Detail	Labor	Matls.	Other	Total cost
PURC			LYOS	1110	FRAMEWORK	30,000	30,000	20,000	80,000
			LYOS	1120	HOUSING	20,000	20,000	30,000	70,000
								00	150,000

of pages 234

approval

PROJECT DEFINITION SUMMARY FORM | Activity no. 1200 | Project no. 1000 Hemomat

Project, breakdown item name	Major components	Timing	
BLOOD SYSTEM	1. COLLECTION SYSTEM	Sched. start 04/26/98	Sched. finish 02/03/99
	2. BLOOD CHEMISTRY	Early start	Late start
	3. DATA I/O	Early finish	Late finish
	4. ACCOMMODATION SYSTEM	Duration	Slack
WBS no. 1200 I-J No.	5.	Paid time	Actual time

Responsibilities				Activity description		Resource allocation			
Direct	Suprtg.	Montrg.	Supplg.	Line no.	Detail	Labor	Matls.	Other	Total cost
DESE	MNFG	QAQC	var	1210	COLLECTION SYSTEM	50,000	100,000	150,000	300,000
	PURCH			1220	BLOOD CHEMISTRY	200,000	150,000	350,000	700,000
	REGL			1230	DATA I/O	50,000	50,000	100,000	200,000
	PRDE								
DESE									

First meeting 01/06/98 | Last meetin 04/02/9

Prepared by Aurora Schimmel | Mana Chi

From *Practical Project Managemen*

PROJECT DEFINITION SUMMARY FORM | Activity no. 1300 | Project no. 1000 Hemomat

Project, breakdown item name	Major components	Timing	
COMPUTER SYSTEM	1. HARDWARE	Sched. start 02/03/99	Sched. finish 09/10/99
	2. SOFTWARE	Early start	Late start
	3.	Early finish	Late finish
	4.	Duration	Slack
WBS no. 1300 I-J No.	5.	Paid time	Actual time

Responsibilities				Activity description		Resource allocation			
Direct	Suprtg.	Montrg.	Supplg.	Line no.	Detail	Labor	Matls.	Other	Total cost
COMP	DESE	QAQC		1310	HARDWARE	100,000	150,000	350,000	600,000
	PURC			1320	SOFTWARE	400,000	100,000	400,000	900,000
					Attachments				
					References				
					Test specifications				
					Appendices				
COMP						500,000	250,000	750,000	1,500,000

First meeting	Last meeting	Issue date	Effective date	Change no.	Effective date	Page of pages
03/21/98	04/12/98	04/13/98	04/13/98			004 of 234

Prepared by	Management approval	Usufructuary approval	Contributor approval	Contributor approval
Janet Soderstrom	Cemale Attaturk	various	various	

From *Practical Project Management*, Anton K. Dekom, New York: Random House, 1987. ©A.K. Dekom. All rights reserved.

FIGURE 20-5. Integrating the Project Breakdown

From *Practical Project Management,* Anton K. Dekom, New York: Random House, 1987.
©A.K. Dekom. All rights reserved.

nent on all lower level forms. On the way up, the total at any level of indenture should always be *more* than the sum of the cost of the components of the lower level. The reason is that each higher level represents more work than simply bringing up the components. The cost ladder is reflected in Figure 20-6.

Likewise, on the way down from total cost, the sum at each level of indenture must be less than the costs on the previous higher level. The system becomes airtight, so to speak:

- Each item of cost is backed in a detail sheet and on a project definition summary form.
- The date for the project and each activity governs the expenditure and scheduled use of resources.
- Each activity has an established predecessor and successor.

CONCLUSION

With this hierarchy of forms and data we can now show

- The job and its activities
- The scheduled start and finish of project and each activity
- The cost of activities, down to line items
- The cost by major category
- Responsibility for each activity and line item by degree
- Members of the team (using the crew sheet) and their schedule.

TASK	Labor Material Other
ACTIVITY	Sum of tasks Assembly Integration Testing
KEY PHASE	Sum of activities Integration Acceptance testing Burden Packaging Spares and field service
PROJECT	Sum of key phases Overhead Profit

FIGURE 20-6.
Layers of Project Costs

21

CHAPTER

PEOPLE IN PROJECTS

People, most often organized in teams, carry out projects. Teams can be hierarchically related, cross-functional (matrix), or a blend of these two. Whether a team exists or not depends on the size and nature of the project. All teams need a leader. Our first concern should be the leader, who in some small projects may also be the only team member.

A TEAM OF ONE

If the project is small enough, the team leader and the entire team might be one and the same person. Industrial engineering cases are good examples. Pat, the Vice President of Manufacturing, says to our engineer, "Find out why we have so much aluminum scrap."

Pat finds that the problem is with the aluminum. Pat asks for a metallurgist to help investigate. Since the aluminum is bought from a vendor, Pat asks for help from Purchasing. Now, Pat has two team members, both of them from other departments. She has a cross-functional or matrix team.

As soon as there is a team, we must start with our basic rule: state the task. The task dictates the team.

IT'S THE TASK

An ancient rule of project management says, "Never ask for a resource, not even money, until you know (or have stated) the task." Hard to resist answering, when higher management asks, "Just how much money do you want for this project?" Try not to answer until the tasks are established.

A Team Task Form

Naturally, there is a form for the purpose. Tasks may be stated in brief, referencing appendices and attachments, or more extensively in the space allowed. Figure 21-1 shows examples of task statements. Figure 21-2 shows a team composition form filled out, based on the Hemomat we have used as example.*

The form has columns:

Line no.—This is housekeeping that makes it easier to reference an item.

Profession, skill, craft name—The skill, trade, or profession needed for this task or for part of the task: Research Scientist, Assembler, Programmer, and so forth.

Skill code—For organizations in which skills are also identified by a code. Makes classification, retrieval, and monitoring much easier.**

411 Mechanical Engineer	505 Manufacturing Specialist
486 Systems Engineer	666 QC Inspector
202 Senior Buyer	*and so forth*

Other columns might be needed, as well: the individual contribution expected from each team member, the name of the person who represents the skill; the department (or contractor) supplying

* Even when managing by computer, task and team must be related. Good computer programs provide for the appropriate input, though the screen may look different than our form. In some project management computer programs, tasks cannot be established without first including a list of available resources. Although this puts the cart before the horse, it's done in the belief that without resources, you can't do the work.

**Real codes, someplace, but not necessarily in your organization.

Task No. 314-317, Stability Test	Resources
■ Chemical Research requests ingredients per SOP 37-123. Ingredients are mixed as specified on Form No. 37R002. A sample is cycle tested per SOP 34-002. QC inspects samples as provided in QC procedures and enters findings on the sample tag. If the sample deviates from limits, the test is aborted. Sample and data are returned to Chemical Research for further process.	Research scientist Laboratory assistant Cycle test facility Test lab aide QC inspector
Task No. 123-125, Program Control Module	**Resources**
■ The systems analyst provides the designated programmer with systems specifications. After review for clarity and approach, the programmer writes the necessary code. The application will be debugged by Operations, which will provide the output and its indicated problems.	Systems analyst Programmer Computer (access) Systems specialist
■ After final debugging and acceptance testing, the application is entered into the ProcLib and a priority is assigned.	Standards supervisor
Task No. F121, Field Service Problem	**Resources**
■ Field Service reports indicate that our units cut out under access demand from more than 26 workstations and power requirements greater than 2,800 W. Under our guarantee, restart is at our expense, and frequent cut-outs are affecting sales (see FSR No. 93-0904).	Field service engineer Power systems engineer
■ Study the reason for the cut out. Redesign power control system, make a new control chip, and retrofit installed units.	An entire major project team

FIGURE 21-1. Examples of Task Statements

the resource. When a "bulk" group—bricklayers, assemblers, among others—is expected to contribute work to the project, that is shown as a line item, too, specifying the probable number of people in the group: Programmers, 4.

What should also be shown, however, is the name of the person responsible for the group: the title or name of a manager or contractor representative. This is necessary to satisfy an earlier condition: *assign the task, assign responsibility.* With that name available, we know whom to call when a functional or contractor problem or question arises. Not to mention that the person named should be considered a member of the project team.

Setting Up the Team

In some entities and situations, project managers can pick the team member. In others the assignment is strictly at the discretion of the functional manager (and, of course, the contractor). Lobbying is not forbidden or can be carried on informally. Communication, however, is always necessary:

- Go to the functional manager whose resources are needed.
- Discuss the project, its requirements and conditions.
- Negotiate:
  - From up to down—Conditions expressed by the project manager to the functional manager. The project, its objectives, what it or the activity or task demand from that function and its resource. "We need to redesign the control chip. Whom do you have who...?"
  - From down to up—Conditions expressed by the functional manager to the project manager. What the functional manager or representative asks for, specifies; or conditions, including probable budget, possible schedule; or response to a question. "I think that'll cost you 500 engineering hours."
  - Up/down/up/down—True negotiation between functional and project manager since both may need more data, latitude, and so forth.
- Define the deliverable and its conditions.
- Write it up, documenting what was agreed.
- Follow up and follow through.

This way of structuring the team can be time consuming. And without management support for the project leader, the effort may be futile. But done right and with proper support, these steps produce favorable results.

PROJECT LEADERS

Wishful thinking aside, no team can function without a leader. In projects, the leader may be temporary, gone at the end of the project. It is equally possible that the temporary condition is even shorter, gone at the end of a phase. But while that leader is in charge, certain rules must be obeyed and circumstances met.

Promulgation

Management must give the project manager (leader, coordinator, whatever) public recognition.

Generic. General or generic promulgation is done through an established policy, the policy on delegation. The project leader is given a temporary rank, say, "director with the powers and authority as delegated in Policy No. 01.04." To know what this "director" can do, sign for, approve, all anyone needs to do is look in the policy manual. To make it official, a bulletin is posted:

> Beatrice Goode is appointed to head project Hot Shoe, with the operating rank of a director.

With that standing and those powers, the project is, truly, hot.

Specific. Not every organization has policies and procedures. When, whether for lack of policies or otherwise, a specific promulgation has to be made, it can be done by enumeration. Let's read the bulletin:

> Tony Blech is Coordinator for project Arquebus. The coordinator will report directly to the Vice President of Marketing; is an *ex officio* member of all project planning meetings; will be informed of all project purchases exceeding $12,000; is a voting member of the Make/Buy and Material Review Committees; receives a copy of every official project report; will have access to all relevant accounting information; and prepares the weekly and monthly summary, activity, and cost reports.

Quite an enumeration! How is a listing like this put together, when do the enumerations stop? Depends on the wisdom and needs of management and those of the project and when they are satisfied.

"Coordinator" is normally a wishy-washy title. In the case of Tony Blech—if the above list were true—the listing makes her quite powerful. Even so, note that Tony has no direct authority.

The way this list reads, whatever Tony wants, Tony gets—especially if there was a proper kick-off meeting for purposes of promulgation. On the other hand, this project leader does not seem to have a team. For our purposes, let there be a team.

Team Management

Our team is composed of the people shown earlier in Figure 20-2: functional representatives from Purchasing, Manufacturing, Design, and Quality; labor from Production and outside contractors. In the example, the team consists of at least six people. With each, the leader should have discussed, defined, and secured agreement on

- The contribution expected from the team member, function, or contractor
- The conditions of the work and the deliverable
- Milestones
- Administrative behavior.

Administrative behavior should be detailed in the kick-off or initialization meeting.

Initialization. As early as possible, even if not all team members are known yet (some won't be known, depending on project phases, until much later), there should be a formative, initialization—kick-off—meeting of the known team members.

During the kick-off meeting, the project leader sets the tone. Curiously or otherwise, the tone set in that meeting lasts for as long as that group of people works together, especially if the project leader renews the initialization from time to time.

Agenda. Part of the agenda for this meeting should be

- Introduction of leader and team members
- Rights and duties of the project leader
  - Authority to sign work orders, purchase orders, receiving reports, inspection reports, and so forth
  - Authority delegated to team members
  - Handling exceptions and variances
- Administrative behavior
  - Hours of work, recurring and other meetings, project manners
  - Communication with the leader and among members
  - Written reports from whom, to whom, and when.

The project leader should make sure that

- A regular meeting is held during each cadence period throughout the project's life.

- Whatever written reports are needed, they precede the meetings or are available at the meeting (whichever is desired or necessary).
- Every team member knows how to reach every other team member, leader included.
- Politicking is minimized, preferably not tolerated.

If written reports are not otherwise required, the project leader should ask that a written report be prepared by key team members for a fixed day in each cadence period and original or copy is sent to the project leader.

PROJECT PEOPLE

The emphases throughout this volume, and particularly in this chapter, are on personal contact and on paper work. Communication and documentation are keys to successful project management. These two requirements often conflict with the psychological make-up of people in projects. So although loathing paperwork, the people try to move mountains with memos. And it isn't working.

Hidalgos

The word "hidalgo" has its origins in Spanish history, but its use here is simply a convenience.* Nor, despite the implication of Spanish orthography, is the designation for males only. All start out with one characteristic: they take risks.

Hidalgos have a hard time saying "No" when a project job is offered. They may already have 12 top-priority jobs, but still accept the new one. Because of that, the walking and talking mentioned below are even more necessary. Therefore, the rules are these:

1. Walk it up—Go there, personalize the contact.
2. Talk it up—Secure agreement and commitment in person.

*It's in the dictionary. Not everyone is a hidalgo. Nor, on a scale of 1 to 10, is everyone a 10. In fact, we are all mixes of several categories. With hidalgos one set of traits sticks out; with others, other traits. Nor are hidalgos better than the rest of us; just inhabited by some singular personality traits. And all classification smacks of stereotyping.

3. Write it up—Document the discussion and agreement.
4. Follow up—Don't leave it to the hidalgo; you do it.

People, Paper, and Planning

Many hidalgos tend to be averse to other people. They often are shy, aggressive when cornered, steel themselves when they have to meet others, and therefore postpone, even avoid human contact. If it is absolutely necessary, a memo gets written. E-Mail is absolutely heaven-sent for them!

Hounding the Hidalgo. Project leaders and other managers feel ill-used because they see themselves forced to "hound" their hidalgos for documents, reports, and plans.

Hidalgos also tend—because they really are risk takers—to jump at the work without necessarily planning. "Management doesn't give me enough time." To the extent that the hidalgo's boss also is a hidalgo, that could be true. "How can I plan, when nobody tells me anything." The hidalgo waits for "them" to come, may not figure on going there. And after all, "I know this job. Why do I have to waste time planning?"

And Reports. As for reports, the same aversion holds. "If the boss wants to know how's it going, she'll come and ask me." Each one of these traits has a negative and a positive side. In the case of reports, they will be few, but concise and to the point. And to the extent possible, routine reports that can be expressed in numbers should be done by Accounting and computers.

We need to know, however, what reports are needed to conduct the project successfully.

CHAPTER 22

PROJECT REPORTING

Because projects are temporary, reporting carries a certain urgency. Project reports must be timely to allow for remedial action. Other than that, reports are reports. They can be

- Numeric—Money, hours, whatever is expressed with numbers
- Narrative—This is what happened, this what didn't
- Graph—Charts of all kinds, usually translations of numeric data
- Hybrid—Combinations of any of the above.

FINANCIAL REPORTS

Financial reports are in the main numeric and are of two types:

Needs of the organization—Financial, numeric reports of hours, costs of labor, material, commitments, accruals, payments, cash flow, forecasts

Needs of the project—Ratios, activity cost detail, categorizations and special control arrangements of data.

JOB LISTING		Project: FONGONA MINE EXTENSION						
Number: F9X2662		Period: November 1, 1998 to January 2, 1999				Basis: 3 shifts/day, 5 days/week		
Charge number	Activity	Duration	Earliest Start	Earliest Finish	Latest Start	Latest Finish	Budget	Actual
SIX TO SURFACE SHAFT RAISE								
4856-060	Timber shaft surface to 3rd level	4	11/04/98	11/10/98	01/07/99	01/13/99	16,000	8,300
4856-060	Extend utilities surface to 3rd level	3	11/04/98	11/09/98	01/08/99	01/13/99	13,000	–
4851-033	Build 3rd level plat	1	11/04/98	11/05/98	–	–	24,000	–
4852-032	Drift 75'S 3rd level plat to DM 100	6	11/05/98	11/13/98	–	–	43,000	11,000
4852-032	Cut for MG and air lift	4	11/13/98	11/19/98	–	–	14,000	–
4801-003	Install motor generator set	3	11/19/98	11/24/98	12/04/98	12/09/98	8,500	–
4851-033	Install air lift	1	11/19/98	11/20/98	12/08/98	12/09/98	6,000	2,000
4852-031	Drift 100'N to DM 125	8	11/19/98	12/02/98	–	–	49,200	–
4852-031	Cross-cut for dam	2	12/02/98	12/04/98	–	–	–	4,200
4852-031	Lay track for 3rd level	1	12/04/98	12/07/98	–	–	–	–
4852-031	Take down tram equipment	2	12/07/98	12/09/98	–	–	–	2,200
4852-033	Drift 125' to DM 250	7	12/09/98	12/18/98	–	–	57,000	6,000
4852-034	Switch back 75' to No. 10 raise	5	12/29/98	01/06/99	–	–	40,000	–
SIX LEVEL NORTH								
4854-061	Two stope development raise	58	12/02/98	01/25/99	04/01/99	06/22/99	72,000	–
4854-061	Start loop drift 50' for ore pass	3	11/02/98	11/05/98	07/16/99	07/21/99	41,800	–
4853-061	Cross-cut for dam, 25' (if necessary)	2	11/05/98	11/09/98	07/21/99	07/23/99	(none)	–
4853-061	Cross-cut cap and powder storage	2	11/09/98	11/11/98	07/23/99	07/27/99	18,000	–
4852-061	Strip for MG set (main level changing)	3	11/11/98	11/16/98	07/27/99	07/30/99	59,600	3,200
4852-061	Strip side track for equipment side lining	4	11/20/98	11/23/98	07/30/99	08/05/99	–	–
4852-061	Install side track and switch	1	11/20/98	11/23/98	08/09/99	08/10/99	–	–
4852-061	Complete loop drift to ore pass	17	11/25/98	12/21/98	08/10/99	09/02/99	–	2,000

FIGURE 22-2. Job Listing

NOTE: Charge numbers are from a CPM chart to which Chart of Account numbers (i.e., -061, -033, etc.) have been added. Considering the latest start and finish dates of Six Level North work, a major scheduling contingency is built into this project.

don't know what you want." Sometimes true. We must specify what we want. There should, moreover, be a common format to each report of a kind. Makes reading and understanding easier.

Don't . . . shoot the breeze. "We are now assembling data for the next activity, which is being considered for a scope change." Huh? Nor should they contain the primal scream. "Nobody is on schedule and the project is going down the tubes!" Maybe true, but how do you know?

Do . . . focus on the items at variance, even if the variance is positive. Recounting everything that went on during the cadence period may make the report boring, even useless. The concentration should be on what did not happen or what exceeded plan.

Thus narrative reports should consist of

- An eye-catcher section—The lead part of the report in which one (or at most, two) items are mentioned; these are the one or two items the reporter wants read if nothing else is read.
- Variances—What is below or over plan, or what exceptional event should be mentioned.
- Closure—What about it? What will the reporter do, or what should who do about it?

Each team member should be required to produce a narrative report each cadence period. A sample of such a report is in Figure 22-3.

Logs

Project logs are a particular type of narrative report. Logs are communication tools particularly needed when various people come and go with their individual work, individual problems, questions, and interface needs. Project leaders should insist that each team member first "tell it to the log." The log, conversely, should be required reading for each team member coming on duty.

Log layouts can be simple, with entries as shown below:

- Activity affected—What are you writing about?
- Variance—What is so important that it had to be written down?
- Closure—What will you do about it? Who else should do something about it?

WEEKLY REPORT	Project: Lab Renovation #3622
Week: 37	Prepared by: David Klein, C.E.

SUMMARY

227-229, THERMAL GLASS INSTALLATION

This activity was completed 3 weeks behind schedule at additional cost of $4340 (12% overrun). Showcase installation, Activity 229-231, is now proceeding and should finish on schedule.

VARIANCES

165-167, AIR EXCHANGE SYSTEM

We were fortunate in that the system was installed and accepted by the environmental inspector within the scheduled time. There was a 25% contingency budget, which can now be reversed. Accounting has already been notified.

167-169, PURIFICATION SYSTEM SHUNT

Thanks to the acceptance of the exchange system as originally designed, the purification system shunt will be eliminated.

300-302, FRONT END LAYOUT

Architecture has not been able to produce an acceptable layout. The next promise date is Week 40, which will adversely affect Activity 308-310, Equipment relocation and installation.

FIGURE 22-3. Example of a Narrative Report

- Attention—Who else should know and/or be told about it?
- Date—When was the entry made?

Figure 22-4 shows a sample log format. Logs should be in bound books, never loose-leaf. That way, nobody will be tempted to take the page along. And if it is taken, the bound book will show it missing.

Distribution. Log entries should be copied and the copies distributed as follows:

1. Activity file
2. Open items file
3. Originator
4. To the attention of whom the entry is addressed (if anyone).

Log entries should be reprised in the cadence narrative report, if the entry is either important enough to mention or not yet closed. In large projects and projects where security (including

PROJECT LOG	Activity	Number
Entry made by (name, department, phone)		Date of entry
Describe variance, problem, finding: ☐ I will handle it. ☐ To the attention of		
Describe what needs to be done: By what date?		
Anything else in this matter:		

FIGURE 22-4. A Project Log Format

NOTE: This image can be imprinted with a rubber stamp on a bound book with ruled pages.

patentability) matters, team members may be asked to keep the equivalent of a personal daily log: a project diary.

Logs and diaries come in handy, besides all that has been already written, when report time comes around. Nothing like having notes as source material for the report or for meetings.

MEETINGS

Meetings can be formal and informal, can involve the team only or the whole complement. Meeting rules are common. One type of meeting must be mentioned. This meeting should include only team members, including contractor representatives, should be managed, but informal, and should allow people to tell the truth to the project leader and one another without judicial proceedings. These team meetings should be on the same day that team members are reviewing project progress—as seen by their expertise and viewpoint, of course. That way the information is current and variances can be handled, hopefully, in time for remedial action.

Part of the speed necessary for that remedial action can also be supplied by computers.

CHAPTER 23

COMPUTERS IN PROJECT MANAGEMENT

Computerized project monitoring systems (CPMS) abound. What anyone uses depends on the amount of data manipulation that might be required, the level of detail we might need, and the response speed desired.

CPMS

CPMSs exist now for any size of project and any size of computer—from simple systems, which make Gantt charts and use basic input for basic reports, to complex mainframe programs, which do almost everything under the sun.

The Positives. To get the best CPMS for a specific project, project people must

- Measure the project to find out what monitoring system it can afford. (Unless there is a substantially idle mainframe computer sitting around, which happens to have a $45,000 CPMS waiting to work, it may be best to look at something smaller. Not to mention that few projects need that much processing power and speed.)

- Determine what really needs to be handled by computer:
  - Make a list of features that are absolutely needed (scheduling charts, for example, Gantt and PERT/CPM).
  - Will changes be frequent or few? If few, do you want a CPMS?
  - Enumerate the reports required and whether they will be repetitive or individual (if individual, look for a database system).
- List the needed items and match them against off-the-shelf CPMSs.
  - If you are contemplating a small system (TimeLine and Workbench are two of the many small ones), buy a demo disk. If you buy the system, get your demo money back.
  - If you're thinking about a large system (Artemis is one), ask for a demo using some of your input; the demo should be free.
  - For an in-between system (Primavera or View Point are examples), ask for the salesperson to come to your site and show how the features relate to your project.

The system that meets your project's needs most closely is for you. If the salesperson promises to fit the system to your needs after you buy it, don't buy it. What isn't ready on Day 1 will cost you more than money.

The Negatives. No system may fully match what a specific project needs; acquiring one calls for some compromise. The compromise should never be more than minor. One of the big negatives is that any CPMS forces a static discipline. Many people, not just the hidalgo type, feel that discipline stifles creativity. True, somewhat, but creativity must be limited anyway, and discipline cannot deteriorate into mindless dictatorship.

CPMSs have a significant negative, invisible costs:

- Learning them—It takes time, and a lot of people prefer to learn by playing around, rather than reading the manual.
- Original data input—Virtually every datum in the CPMS must be input by a human. This takes time, money, and the inclination to do so. (Hidalgo types neither want to study the manual, nor "waste my time" at such clerical work. "I'm busy doing *real* work.")

- Updates—As changes occur, somebody, "but not me," has to post them into the computer and its database.

It is unavoidable, but we must remember that once the data are loaded, normally all that is required is to update and enjoy the output. Not to mention that the output will be accurate, easy to obtain, nicely done, and legible. Besides, there are ways to make life with your CPMS easier.

EASIER USE

If the CPMS is for a mainframe, the vendor supplies training, manuals, blank forms, and a host of other items to justify the price and needs of the program. This section is for those who buy a system for $5,000 or less, where the price may not justify all the bundled freebies.

Read the Book

One person should be responsible for knowing what the CPMS does, wants, and how it does what you want. This person will be the Standards Specialist, who actually has read the manual that came with the CPMS and knows the program inside and out.

Defaults. "Default" is computereze for what the CPMS (or any program) does if nobody says differently. The Standards Specialist should know and prepare a list of program defaults. Most CPMSs default, for example, to a black-and-white screen if users do not specify a color terminal.

Virtually all CPMSs default to the cadence of a 40-hour week that begins on Monday and ends on Friday. For the majority of projects this will be all right. If not, someone must specify the cadence to the CPMS.

Musts. Musts are things the CPMS demands so that it can do something else. One of the CPMS's most common musts is that you must enter the resources before you can enter the activity. The logic is self-evident: no resources, no job. Other musts involve duration and the cadence. If the cadence is in days, the CPMS will not accept an activity in hours.

A key must is in CPMSs that are written in CPM (Critical Path Method) notation. The manual may not even say so, but there must be an activity called Start and one called Finish (Stop or End). Start and Finish consume no time nor other resources. Without them, however, the CPMS will not allow correct activity input.

In the category of musts are hidden rules and actions. In at least one CPMS, rogue activities (those without either a Start or a Finish) can be added without apparent penalty. But the program will add the time value of that activity to the critical path without telling you.

Find It

Those "hidden" rules can be discovered by having someone put in a short, 8- to 10-activity, practice network, the calculations for which have been done manually beforehand. The critical path time of this network should also be precalculated so you can compare it with the finding of the CPMS.

In most commercial software packages used in a CPMS, some things are not clear in the hard copy manual or (if available) the on-screen tutorial. Some are discovered by the user, who then must call the software publisher's hotline. Some are discovered by exchanging ideas through a software users' group. In all cases, the information must be clearly documented and made available to all the software users in your organization. This can be done through an in-house users' bulletin maintained by the Standards Specialist or, in a larger organization, by the computer resources center. Such a bulletin board can also be used to pass along updates received from the software publisher and ideas developed internally.

Screens and Forms

Another necessary job is to

- Cull all the forms and formats from the manual.
- Look at the screen for the same and other forms and formats.
- Make paper samples of these forms and formats.

Samples should be made per type or category. Activity input is a case in point. Somebody has to cull, look, and make a form that shows everything (or key items) about, for example, the activity.

Activity Formats. What information will the CPMS ask about an activity or about resources or budget data? Life will be a lot easier if all related information about an item be on one format, form, sheet, or screen. Figure 23-1 is a short example.

Two aspects of the CPMS are shown on the example. One is the information that the CPMS actually asks about the task; for example, the rules for name of the task, priority, resources.* These the CPMS asks and needs. Then there are explanations prepared by our Standards Specialist:

- "For the computer, the task name must be 21 characters or fewer."
- "End" [Date]—That's what the CPMS wants, if a predetermined end date must be entered.
- "Please stay with the same convention for dates. E.g., 10/22/99."

Abbreviations. "For the computer," we found out, "the task name must be 21 characters or fewer." The CPMS offers us only 21 characters. What to do about tasks that exceed this limit? In the resource listing, some CPMSs allow only 8, some 10 characters for a resource name or designation. Almost any skill is longer than that. Mechanical Engineer takes 19 characters, but only 10 are allowed.

Procedures. What needs doing is simple, but tedious, and calls for much coordination. Somebody, our Standards Specialist,** might have to do more work—come up with a project nomenclature and standard abbreviations—and publish them from time to time. Figure 23-2 is an example of a format for skill abbreviations. A similar format should be used for activity names, even if we can assume that everybody might use "Assy" for Assembly, "Insp" for Inspection, and so forth.

Nomenclature and abbreviations are needed by all concerned and should be updated from time to time. On the other side of all this, team members should be made aware that they cannot by themselves invent new words and abbreviations, but that the need

---

* Note the hidden rule on the form: "Can be used to resequence. Will, however, override normal scheduling." Some large organizations have software resource specialists on staff to assist in these matters.

**Only large projects and enlightened organizations have Standards Specialists. In most cases the project manager or a passionate volunteer does this work.

TASK ENTRY FORM *Use a separate form for each task.*

Name of task for computer entry—21 characters or fewer.

| | | | | | | | | | | | | | | | | | | | | |

Full name of task for your records. ◄ *This data will not compute.*

Type of task: ☐ fixed ☐ ASAP ☐ ALAP ☐ span ◄ *Choose one.*

Start date: Plus [ ] minutes, hours, days, weeks, months. ◄ *Format optional.*

End date: Plus [ ] minutes, hours, days, weeks, months. ◄ *Format optional.*

Time to complete: ◄ *Field cannot be accessed by user. Entered by computer, based on your timing input.*

Current status: ☐ Future ☐ Started ☐ Done ◄ *Choose one, and the computer will tell you how your choice affects scheduling.*

Notes: ◄ *To insert miscellaneous material, as a scratch pad, to resequence tasks, to search key words within a field.*

Priority: ☐ Extra low ☐ Low ☐ Medium ☐ High ☐ Very high ◄ *Choose one. Can be used to resequence. Will override normal scheduling.*

Summarizes schedule named: ◄ *Goes to subnetworks.*

RESOURCES USED ◄ *The computer form shows 2 sets of these columns. Each pair is 10 spaces wide.*

Who/Cost?	Amount	Who/Cost?	Amount

©1993 by A.K. Dekom. All rights reserved.

FIGURE 23-1. A Basic Project Set-Up Form

ABBREVIATED NAMES	Project name	Project number	Date created by
Line number	Short name	Regular name (use more than 1 line if necessary)	Reference
	/ / / / / / / / /		
	/ / / / / / / / /		
	/ / / / / / / / /		
	/ / / / / / / / /		
	/ / / / / / / / /		
	/ / / / / / / / /		
	/ / / / / / / / /		
	/ / / / / / / / /		
	/ / / / / / / / /		
	/ / / / / / / / /		
	/ / / / / / / / /		
	/ / / / / / / / /		
	/ / / / / / / / /		
	/ / / / / / / / /		
	/ / / / / / / / /		
Date entered by		Hard copy distribution	Page of pages

FIGURE 23-2. Abbreviations List Format

From *Practical Project Management,* Anton K. Dekom, New York: Random House, 1987, p. 414. © 1987 A.K. Dekom. All rights reserved.

for a new designation should be checked with a specific person: the Standards Specialist, of course.

ALWAYS PAPER

The manual has been rewritten for the people who really need to know what is involved in using the CPMS. Nomenclature and dictionary documents have been set up. Musts and defaults have been made known. Now we can introduce the computer into the project.

Most CPMSs have rules for that, too, so we follow them to set the project up in the computer—everybody else has already approved it. All the team members are ready to enter activities, data, and so forth. They shouldn't!

Few will follow the next suggestion; here it is, anyway. *Nobody may make an entry into the CPMS unless they first write the entry on paper.* If it isn't written, look out for these pitfalls:

- Only the person creating the entry will know about it.
- The entry may add to or modify existing data, and nobody will know it.
- The entry may change conditions, and only one person will know why.
- No audit trail will exist to reconstruct the changes.

Not to mention that if the entry is in the head of only one of the precious skills on the team, nobody else, such as a clerk, can help.

Anyway, entries have been made. The project is on the books, in the computer, and in process. Milestones will be met with ease or difficulty, costs and specifications may have to be changed and recast, and hair will turn gray.

The project now marches inexorably to its conclusion. For the project leader and the team, the conclusion is a desired moment, but involves more work.

24 CHAPTER PROJECT CONCLUSION

Few projects are aborted. They usually go on to their conclusion, though the road may have turned overland, the cost over budget, and the schedule overrun. Most projects, however, are within budget, on or close to schedule, and produce the fruit expected by all. Chiefly for these reasons, managing and participating in projects is fodder for a successful career.

BRINGING IN THE SHEAVES

The project has been going on:

- The approval system has been used, obeyed, and the project was started with the blessing of higher management and the client.
- The project has been dissected by the usual planning methods; tasks have been found, defined, and assigned.
- Budgets and schedules were made and milestones established.
- Reports were issued, meetings and other reviews held in time for remedial action.

- Corrective actions were undertaken and documented, and change procedures followed our progress.

The project is coming to its expected end. Here comes the client to check it out.

The Closing

Despite how it may seem, the number of ways to close a project is limited. Projects can close milestone by milestone, in one hand-over (or turnkey) parade, or both ways. Evidently projects can also close hodge-podge, with everybody trampling over the grass.

This, however, is not desirable. Where trampling occurs,

- The project leader and team are never in the clear.
- Any fault can be blamed on the leader and team.
- The inevitable schedule and budget overruns become the liability of the leader and team.

Don't let yourself in for it. Of course, an orderly closure is similar to good project planning. People say it takes extra time they don't have, and "whatever needs fixing will be handled later." By whom?

Integration. If a project is handed over milestone by milestone, accepted or "closed" milestones must not be allowed to re-open. A closed milestone should not be questioned later, if for no other reason than if one closed milestone can be re-opened, all of them can.

One problem is bad interface between milestones. Yes, activity 123 does meet its standards, and so does activity 456, but they don't fit with one another. This can be a big problem, but need not be the responsibility of the teams that produced those activities.

To fix the integration problem, start with the standards worked out at the beginning of the project. We can be held responsible only for what we promised; the rest is someone else's problem. Do not allow others to give you their problems.

Then address the integration problem by working out a subsidiary plan:

- Determine the interface problem or need—What must be done to make or ensure the interface?

- Prepare a task statement:
  - Specify the condition that must be met, achieved, or produced.
  - Figure its cost or budget.
  - Add its time value to the schedule.
  - Publish a change order with all the necessary approvals.
- Enter the new task into the system.
- Do the new task.

If the integration problem arises from faulty project work, the same procedure applies, but the project leader and somebody on the team has to own up to it.

Check-Off

A check-off or acceptance form must be created and used for project conclusion. The form should

- Name and describe the activity being handed over
- The activity's conditions of performance or acceptance
- Budget and schedule data *vs* actual data
- Change orders, if any, that affected any of the above
- Signatures of those proffering and those accepting.

Grand Parade

Turnkey projects are handed over in what may be called a grand parade: a handing over of the project in one fell swoop. Size of the project does not seem to matter. What does matter is how to keep track of the inevitable variances between what is being delivered and what has been promised or expected. For the purpose, we need a checklist. The construction business has given this recitation of variances the name "punch list."

Punch List. Preprinted punch lists may be used where the routine is well understood, as on construction jobs since those projects are relatively ritualized. Such punch lists—or adjustment sheets—show the key items of the project by class or group and with subsets of the class or group. Plumbing may be a group. Within plumbing may be subgroups for water, waste, and drainage, for sanitary fixtures, for the fountain in front of the building that will aerate and cool the air conditioning water, and for whatever else.

For other projects where the turnkey job is not so standardized, an open punch list may be needed (Figure 24-1). The list asks

- What is at variance?
- Identify the variance.
- To whom is the variance attributable?
- Distribute the cost of the variance.
- Signatures.

Signatures indicate that the cost charge, who pays for what, is agreed.

Charges. The project team is not always the cause of a variance. The variance could occur because the client changed its mind and the agreed-to standards aren't what they expected. "Now that I see it, I don't like it." As a result, the cost (and, of course, the time) of adjusting the variance—if, in fact, it can be adjusted—must be someone's responsibility. Three possibilities:

1. We didn't deliver what we promised. Our fault, our cost.
2. We did, but you want something different. Your fault, your cost.
3. What's wrong is difficult to attribute. Let's split the cost.

In the form shown, "owner" is the client and "contractor" is the project team's organization(s).

Detail. The open punch list, or adjustment sheet, must be backed by detail. "Repair cam gear" is a good enough summary entry, but may need a lot more explanation and detail for all to realize that the "cam gear has to be re-tooled because the eccentric of the centerpin is showing excessive wear . . . and a change in drawing no. 99-123 . . . ." Back-up data should be referenced right on the adjustment sheet. Might take a wide sheet!

If the adjustments are major, an entirely new project might have to be written up, and we begin . . . at the end. We are, however, coming to the real end.

ADJUSTMENT SHEET		Project	Number	Meeting date	Sheet	of	sheets
Line no.	Variance identification		Cost allocation			Acceptance by	
	Number	Variance and explanation	Owner	Contr.	Joint	Owner	Contr.
Owner representative			Contractor representative			Date	

FIGURE 24-1. A Punch List

THE END

Ending preparations are as necessary and critical as starting preparations. A checklist with the key steps involved in project closure is shown in Figure 24-2. The list starts with the people on the team.

Team Members. One of the key worries is the project team. Team members must be returned to the provider. And they must be returned before they become redundant, obsolete, and a burden to themselves and the project. If they charge time to the project after their useful contributions are made, their cost will be an overrun.

Equipment. Equipment also must be let go at the end of the use period. If the equipment is owned by the entity, the cost of holding on beyond scheduled time may not be visible. If the equipment is rented or leased, holding on may show up as extra cost. It will show up as major extra cost if the equipment is major: cranes, computers, other heavy or high-tech stuff.

Parts. What is more difficult to find: a part meeting Engineering Change No. 2 or one meeting Engineering Change No. 222? Certainly the earlier one. It is a wise policy to inventory spare parts left over from the evolution of the project and make them available to others who will need them.

On construction jobs it is customary to save extra tiles, partly used cans of paint, and so forth. In some projects regulated by public law, "spares" may be part of the history of the fruit: how did this product evolve? Earlier versions, components, and so forth *must* be documented and held for whatever statutory period may be prescribed by law, regulation, or the organization's own proof for its quality, GMP (good manufacturing practice), GLP (good laboratory practice), or other similar files.

Paperwork. Documentation must be cleaned up, properly filed, inventoried, and handed over to the client. If nobody else, the Internal Revenue Service may be interested in the evolution and cost of the project. Most often, however, project documentation contains evolutionary history, tryouts, alternatives tried and discarded, operating information and other valuable data that should be preserved in an orderly manner and formally handed over. Get a receipt.

PEOPLE	■ Release them at the end of their assignments—*formally,* not casually. ■ If on the project for some time, *praise,* even "kick" if necessary.
EQUIPMENT	■ Return, release.
PARTS, SPARES	■ Inventory, document, save. ■ Hand over to Operations, Field service.
SALVAGE, SCRAP	■ Handle salvage like spares. ■ Dispose of scrap for profit.
DOCUMENTATION	■ Clean up, inventory, hand over; if nothing else, Accounting needs it.
HISTORY	■ Write a glorious narrative—it *is* career fodder.

FIGURE 24-2. Closing Down the Project

Glorious History. Last, but not least, write a narrative report describing how gloriously the project team managed to do this job. Earlier it was stated that project work is career fodder. Well, it is. Some warriors do die in battle, but more survive. They get medals—in business, bonuses and stock options—and promotions. Today's Field Marshal might have been yesterday's project programmer.

As an erstwhile corporal, Napoleon Bonaparte, reputedly said,

> *Chaque soldat porte dans sa sacoche un baton de maréchal.*
> Each soldier carries in his bag the baton of a marshal.

■ ■ ■ ■ ■

INDEX

WITHDRAWN
Bridgeport
Public Library
1230066690